PRAISE FOR MICHAEL BARTON

From "*The Power of Regenerative Agriculture:*"

This is your go-to book to help inspire and educate yourself about the possibility of how you can play a role in the interconnection of farming systems with the entirety of the ecological system.

— D. G. FARNSWORTH

Well done book. I liked the chapter summaries at the end of each chapter. Chapter 6, was spot on in describing the challenges and barriers to implementing regenerative agriculture. The author has a passion for this subject and has provided very clear and concise data. The Afterword ends the book on a positive note and a view of hope that we'll get there.

— SRP

These guides to being environmentally responsible often come across as a little preachy. This is not one of those. It's informative and well-written, with a cheerful attitude toward ecologically friendly farming methods.

— LWHIT

AGROFORESTRY AND SILVOPASTURE

AGROFORESTRY AND SILVOPASTURE

A BEGINNER'S GUIDE TO UNITING LIVESTOCK, TREES, AND COMMUNITIES TO COMBAT CLIMATE CHANGE AND BOOST BIODIVERSITY

SUSTAINABLE AGRICULTURE

MICHAEL BARTON

Book Bound Studios

The information contained in this book is based on the author's personal experiences and research. While every effort has been made to ensure the accuracy of the information presented, the author and publisher cannot be held responsible for any errors or omissions.

This book is intended for general informational purposes only and is not a substitute for professional medical, legal, or financial advice. If you have specific questions about any medical, legal, or financial matters, you should consult with a qualified healthcare professional, attorney, or financial advisor.

Book Bound Studios is not affiliated with any product or vendor mentioned in this book. The views expressed in this book are those of the author and do not necessarily reflect the views of Book Bound Studios.

This book is dedicated to the passionate farmers, foresters, and researchers who have embraced the harmonious union of agriculture and forestry, paving the way for a sustainable future. Your unwavering commitment to nurturing the land, preserving our planet's resources, and sharing your knowledge has inspired the creation of this humble work.

In nature, nothing exists alone.

— RACHEL CARSON

CONTENTS

INTRODUCTION

In today's rapidly changing world, the need for sustainable and efficient agricultural practices has never been more critical. With a growing global population and increasing pressure on our planet's finite resources, it is essential to explore innovative solutions that can help meet food, fuel, and fiber demands while preserving the environment. Agroforestry and silvopasture are two such solutions that have the potential to revolutionize the way we approach agriculture and land management. This chapter will explore the fascinating world of agroforestry and silvopasture, exploring their origins, principles, and benefits.

Agroforestry, at its core, is the integration of trees and shrubs into agricultural systems. This practice combines the best of both worlds: the productivity and efficiency of agriculture with the ecological benefits of forests. By incorporating trees into farming systems, agroforestry can provide many benefits, including improved soil fertility, increased biodiversity, enhanced water quality, and reduced erosion. Moreover, agroforestry systems can also provide valuable products such as timber, fruits, nuts, and medicinal plants, diversifying farmers' income sources and contributing to food security.

Silvopasture, a subset of agroforestry, is the practice of integrating trees, livestock, and forage production on the same land. This system offers a unique opportunity to maximize the use of available resources, as trees provide shade and shelter for livestock, while their roots help to maintain soil structure and prevent erosion. In turn, livestock can help manage vegetation, fertilize the soil, and control pests. The combination of these elements creates a dynamic, resilient, and productive ecosystem that can support both agricultural and environmental goals.

The concept of agroforestry and silvopasture is not new; in fact, it has been practiced for centuries by indigenous communities around the world. However, the modern agricultural revolution, with its focus on monoculture and intensive farming practices, has led to a decline in the use of these traditional systems. As we face the challenges of climate change, resource scarcity, and environmental degradation, there is a renewed interest in agroforestry and silvopasture as sustainable alternatives to conventional agriculture.

In the following sections, we will explore the historical context of agroforestry and silvopasture, delving into their origins and evolution over time. We will then examine the fundamental principles and components of agroforestry systems, highlighting the various techniques and strategies that can be employed to create diverse, resilient, and productive landscapes. Next, we will focus on silvopasture, discussing its benefits and techniques in greater detail. Finally, we will present case studies of successful agroforestry and silvopasture implementations from around the world, showcasing the potential of these systems to transform agriculture and contribute to a more sustainable future.

Join us on this exciting journey as we uncover the potential of agroforestry and silvopasture and discover how these innovative practices can help to shape the future of sustainable agriculture.

A Brief Historical Context of Agroforestry and Silvopasture

The roots of agroforestry and silvopasture can be traced back thousands of years, with evidence of these practices found in various ancient civilizations across the globe. To fully appreciate the significance of these sustainable agricultural systems, it is essential to delve into their historical context and understand how they have evolved.

Agroforestry, the intentional integration of trees and shrubs into crop and animal farming systems, has been practiced for millennia. Ancient civilizations, such as the Maya in Central America, the indigenous people of the Amazon Basin, and various African tribes, recognized the benefits of combining trees with agriculture. These early societies understood that trees played a crucial role in maintaining soil fertility, preventing erosion, and providing shade and habitat for wildlife. They also valued the diverse array of products that trees could provide, such as fruits, nuts, timber, and medicine.

Similarly, silvopasture, which involves the integration of trees, livestock, and forage plants, has a long and rich history. In Europe, silvopasture can be traced back to the Roman Empire. During this period, farmers grazed their livestock in wooded areas to take advantage of the natural forage and shelter provided by the trees. This practice continued throughout the Middle Ages and is still common in many parts of Europe today.

In Asia, traditional agroforestry systems, such as the home gardens of Indonesia and the tree-based agriculture of the Philippines, have been practiced for centuries. These systems not only provided food and income for local communities but also played a vital role in preserving biodiversity and maintaining ecological balance.

Despite their ancient origins, agroforestry and silvopasture experienced a decline in popularity during the *20th century,* as modern agricultural practices prioritized monoculture and the use of synthetic fertilizers and pesticides. However, in recent decades, there has been a resurgence of interest in these sustainable agricultural systems, driven by a growing awareness of their environmental, social, and economic benefits.

Today, agroforestry and silvopasture are recognized as valuable tools in the quest for sustainable agriculture. Numerous organizations, researchers, and farmers are working to promote adoption and further refine their techniques. By understanding the historical context of these practices, we can appreciate their enduring relevance and potential to shape the future of agriculture in a more sustainable and resilient direction.

Key Principles and Components of Agroforestry Systems

Agroforestry, a sustainable land management practice, combines agriculture and forestry to create integrated, diverse, productive, and resilient ecosystems. This innovative approach to land use offers numerous benefits, including increased biodiversity, improved soil health, and enhanced economic opportunities for farmers. This section will delve into the fundamental principles and components of successful agroforestry systems.

Diversity: Agroforestry systems thrive on diversity, both in terms of plant species and agricultural practices. By incorporating a variety of trees, shrubs, and crops, these systems create a more resilient ecosystem that can better withstand pests, diseases, and climate fluctuations. This diversity also provides a range of products and services, such as food, timber, and habitat for wildlife, which can contribute to the overall sustainability and profitability of the system.

Integration: The integration of trees, crops, and livestock is a fundamental aspect of agroforestry systems. This integration allows for the efficient use of resources, such as sunlight, water, and nutrients, as well as the recycling of organic matter. Trees can provide shade and shelter for crops and livestock, while their roots help to improve soil structure and prevent erosion. In turn, crops and livestock can contribute to the system's overall productivity by providing food, fiber, and other valuable products.

Synergy: Agroforestry systems are designed to maximize the positive interactions between trees, crops, and livestock. These synergies can increase productivity, improve resource use efficiency, and enhance ecosystem services. For example, trees can help to improve soil fertility by fixing nitrogen, while livestock can help to control weeds and pests through grazing. By working together, the various components of an agroforestry system can create a more sustainable and productive landscape.

Adaptability: Agroforestry systems are inherently adaptable, as they can be tailored to suit various environmental conditions, cultural practices, and economic objectives. This adaptability allows farmers to experiment with different tree species, crop rotations, and livestock management strategies to find the best fit for their unique circumstances. By being flexible and responsive to change, agroforestry systems can help to build resilience in the face of climate change and other environmental challenges.

Sustainability: The ultimate goal of agroforestry is to create sustainable agricultural systems that provide long-term benefits for both people and the environment. By promoting biodiversity, conserving natural resources, and enhancing ecosystem services, agroforestry systems can contribute to the overall health and resilience of our landscapes. Additionally, by diversifying farm income and providing new market opportunities, agroforestry can help to support the economic viability of rural communities.

In conclusion, the fundamental principles and components of agroforestry systems – diversity, integration, synergy, adaptability, and sustainability – work together to create a holistic approach to land management that benefits both people and the environment. By understanding and applying these principles, farmers, landowners, and policymakers can help to promote the widespread adoption of agroforestry and contribute to a more sustainable future for agriculture.

Understanding Silvopasture: Benefits and Techniques

Silvopasture, a vital component of agroforestry, is a sustainable land management system that combines the cultivation of trees, forage plants, and livestock in a mutually beneficial environment. This innovative approach to agriculture offers numerous advantages for both the ecosystem and the farmers who practice it. In this section, we will delve into the benefits of silvopasture and explore the techniques used to create a successful silvopastoral system.

Environmental Benefits: Silvopasture systems contribute to the preservation of biodiversity, soil health, and water quality. Integrating trees and forage plants helps prevent soil erosion, improve nutrient cycling, and reduce water runoff. Additionally, trees provide habitats for various species of birds, insects, and other wildlife, promoting a more diverse and balanced ecosystem.

Climate Change Mitigation: Trees in silvopasture systems act as carbon sinks, absorbing and storing atmospheric carbon dioxide, which helps to mitigate the effects of climate change. Moreover, the shade provided by trees reduces the need for livestock to expend energy on regulating body temperature, resulting in lower animal methane emissions.

Economic Benefits: Silvopasture systems can provide multiple income streams for farmers, as they can profit from selling timber, non-timber forest products, and livestock. Diversifying income sources reduces the financial risks associated with traditional monoculture farming and can increase overall profitability.

Improved Animal Welfare: Trees in silvopasture systems offer shelter and shade for livestock, protecting them from harsh weather conditions and reducing heat stress. This improved environment results in healthier animals, which can lead to higher productivity and better-quality meat, milk, and other animal products.

Techniques for Implementing Silvopasture

Tree Selection: Choosing the right tree species is crucial for a successful silvopasture system. Some factors to consider are the tree's growth rate, shade tolerance, and compatibility with the forage plants and livestock in the system. Native tree species are often preferred, as they are well-adapted to local conditions and can provide additional benefits for wildlife.

Planting and Spacing: The arrangement and density of trees in a silvopasture system will depend on the farmer's specific goals and the livestock's needs. Trees can be planted in rows, clusters, or scattered throughout the pasture. Proper spacing is essential to ensure adequate sunlight for forage plants and facilitate livestock movement.

Forage Management: Silvopasture systems require careful management of forage plants to maintain productivity and prevent overgrazing. Rotational grazing, in which livestock are moved between different paddocks or sections of the pasture, can help to promote even grazing and allow forage plants to recover.

Livestock Management: The choice of livestock species and the stocking density will depend on the specific objectives of the silvopasture system and the available resources. It is essential to monitor the health and well-being of the animals and to adjust management practices as needed to ensure their welfare and productivity.

In conclusion, silvopasture offers a promising solution for sustainable agriculture, providing environmental, economic, and animal welfare benefits. By understanding and implementing the appropriate techniques, farmers can create thriving silvopastoral systems that contribute to a more sustainable and resilient agricultural landscape.

Case Studies: Successful Agroforestry and Silvopasture Implementations

In this section, we will delve into real-life examples of agroforestry and silvopasture systems that have been successfully implemented around the world. These case studies showcase the potential of these sustainable agricultural practices to improve the livelihoods of farmers, enhance biodiversity, and contribute to climate change mitigation. Furthermore, by examining these success stories, we can gain valuable insights into the practical application of agroforestry and silvopasture principles and techniques.

Case Study 1: The Vi Agroforestry Project in East Africa

The Vi Agroforestry Project, initiated in 1983, has been working with smallholder farmers in Kenya, Tanzania, Uganda, and Rwanda to promote agroforestry practices. The project has reached over 1.8 million people and helped plant more than 120 million trees. By integrating trees into their farms, the farmers have diversified their income sources, improved soil fertility, and enhanced food security. The project has also contributed to climate change mitigation by sequestering an estimated 2.5 million tons of carbon dioxide.

Case Study 2: Silvopasture in the Dehesa System of Spain and Portugal

The Dehesa system, a traditional land-use practice in Spain and Portugal, is an excellent example of silvopasture. In this system, livestock grazes on grasslands interspersed with oak trees, primarily holm oaks and cork oaks. The trees provide shade and fodder for the animals, while their acorns are a valuable food source for the famous Iberian pigs. The Dehesa system supports a diverse range of flora and fauna, and it has been recognized for its role in preserving the Mediterranean landscape and its cultural heritage.

Case Study 3: The Quesungual Slash and Mulch Agroforestry System in Honduras

The Quesungual Slash and Mulch Agroforestry System (QSMAS) was developed by smallholder farmers in western Honduras to respond to land degradation and food insecurity challenges. This innovative system involves the selective pruning of trees and shrubs, which are then used as mulch to improve soil fertility and moisture retention. The QSMAS has successfully increased crop yields, reduced the need for chemical inputs, and enhanced biodiversity. It has also been recognized for its potential to contribute to climate change adaptation and mitigation.

Case Study 4: Alley Cropping in the United States

Alley cropping is an agroforestry practice that involves planting rows of trees or shrubs alongside annual or perennial crops. In the United States, alley cropping has been successfully implemented in various regions, including the Midwest and the Southeast. For example, in Missouri, farmers have planted rows of black walnut trees with soybeans, corn, and wheat. This system provides multiple benefits, such as improved soil health, increased biodiversity, and additional timber and nut production income.

These case studies demonstrate the versatility and adaptability of agroforestry and silvopasture systems across different regions and ecosystems. By learning from these successful implementations, we can continue to refine and expand the use of these sustainable agricultural practices, paving the way for a more resilient and environmentally friendly future for agriculture.

The Future of Sustainable Agriculture through Agroforestry and Silvopasture

As we conclude our journey through the introduction to agroforestry and silvopasture, it is essential to take a step back and appreciate the immense potential these practices hold for the future of sustainable agriculture. In a world grappling with climate change, environmental degradation, and food security challenges, agroforestry and silvopasture offer innovative solutions that can **revolutionize** the way we approach agriculture and land management.

Throughout this introduction, we have explored the historical context of agroforestry and silvopasture, delved into their fundamental principles and components, and examined the numerous benefits and techniques associated with these practices. We have also looked at real-life case studies that demonstrate the successful implementation of agroforestry and silvopasture systems around the world.

As we look to the future, it is clear that agroforestry and silvopasture have the potential to play a significant role in addressing some of the most pressing challenges of our time. By integrating trees and livestock into agricultural systems, we can enhance biodiversity, improve soil health, increase carbon sequestration, and promote more efficient use of resources. These practices also offer economic benefits to farmers by diversifying income streams, reducing risk, and increasing overall productivity.

However, realizing the full potential of agroforestry and silvopasture will require concerted efforts from various stakeholders, including farmers, researchers, policymakers, and consumers. Investing in research and development is crucial to advance our understanding of these practices and develop innovative techniques that maximize their benefits. Policymakers must create enabling environments that support the adoption of agroforestry and silvopasture through incentives, education, and extension services. Consumers also have a role to play in driving demand for sustainably produced products, thereby encouraging more farmers to adopt these practices.

In conclusion, agroforestry and silvopasture represent a promising pathway toward a more sustainable and resilient agricultural future. By harnessing the power of nature and working in harmony with the environment, we can create food systems that nourish both people and the planet. So, as we move forward, let us embrace the principles of agroforestry and silvopasture and work together to build a brighter, greener future for generations to come.

1

THE HISTORY AND EVOLUTION OF AGROFORESTRY

A futuristic agroforestry landscape, merging the wisdom of traditional practices with cutting-edge technology, to demonstrate the potential of sustainable agriculture in the future.

Agroforestry, the practice of integrating trees and shrubs into agricultural systems, has been a cornerstone of human civilization for thousands of years. As societies evolved and agricultural practices advanced, the benefits of combining trees with traditional farming methods became increasingly apparent. This chapter delves into the rich history of agroforestry, tracing its roots from ancient civilizations to modern innovations and challenges.

The concept of **agroforestry is not new**; in fact, it has been practiced by various cultures across the globe since the dawn of agriculture. Integrating trees and crops or livestock provides numerous benefits, such as improved soil fertility, enhanced biodiversity, and increased resilience against climate change. These advantages have been recognized and harnessed by our ancestors, who developed sophisticated agroforestry systems that laid the foundation for modern practices.

In this chapter, we will explore the historical roots of agroforestry, beginning with ancient civilizations that employed these techniques to sustain their communities and foster environmental stewardship. We will then examine the development of silvopasture systems, which combine forestry and grazing in a mutually beneficial relationship. As we move through the timeline, we will discuss the innovations and challenges that have shaped modern agroforestry and how these practices play a crucial role in sustainable agriculture and climate change mitigation.

By understanding the history and evolution of agroforestry, we can appreciate the wisdom of our ancestors and the importance of continuing to develop and refine these practices for the benefit of future generations. Furthermore, as we face unprecedented environmental challenges, the lessons from the past can serve as a guiding light, illuminating the path toward a sustainable and resilient future for agriculture and our planet.

Ancient Agroforestry Practices and Civilizations

The roots of agroforestry can be traced back to the dawn of human civilization when our ancestors began cultivating crops and domesticating animals. As societies developed, so did their understanding of the complex relationships between plants, animals, and the environment. This section will explore the ancient agroforestry practices and civilizations that laid the foundation for modern agroforestry and silvopasture systems.

One of the earliest examples of agroforestry is found in the indigenous cultures of the Amazon Basin. The Amazonian people practiced a form of agroforestry known as *"forest gardening,"* which involved cultivating multiple layers of vegetation, including trees, shrubs, and herbaceous plants. This diverse and sustainable system provided food, medicine, and other resources for the community while maintaining the integrity of the surrounding ecosystem.

In Africa, agroforestry has been practiced for thousands of years, with evidence of tree cultivation dating back to *6000 BCE.* Ancient African societies recognized the benefits of integrating trees into their agricultural systems, such as improving soil fertility, providing shade for crops, and serving as a source of food and medicine. The practice of planting *Faidherbia albida*, a nitrogen-fixing tree species, alongside cereal crops is a prime example of traditional African agroforestry.

The ancient Mayan civilization in Central America also utilized agroforestry techniques in their agricultural systems. The Mayans cultivated various crops, including maize, beans, squash, and cacao, often intercropping them with fruit and nut trees. This approach maximized the use of available land and promoted biodiversity and soil health.

In Asia, the traditional rice paddy systems of China, Indonesia, and the Philippines incorporated trees and fish into their agricultural landscapes. These integrated systems provided multiple benefits, such as erosion control, nutrient cycling, and increased food production. Similarly, the ancient terraced agroforestry systems of the Andean region in South America combined trees, crops, and livestock to create a sustainable and productive landscape.

As these ancient civilizations demonstrate, agroforestry has been integral to human societies for millennia. The knowledge and techniques developed by our ancestors have been passed down through generations, shaping the foundation of modern agroforestry and silvopasture practices. By understanding and appreciating the historical roots of agroforestry, we can better comprehend its significance in contemporary agriculture and the potential it holds for addressing the challenges of climate change and sustainable food production.

The Development of Silvopasture Systems

Silvopasture, a practice that combines forestry and the grazing of domesticated animals, has a rich and diverse history that has evolved over time. This section will delve into the development of silvopasture systems, exploring their origins, their various forms, and how they have been adapted to suit different environments and cultures.

The roots of silvopasture can be traced back to ancient civilizations, where the integration of trees and livestock was a common practice. In ancient Rome, for example, farmers recognized the benefits of combining trees with pasturelands. The trees provided shade and shelter for the animals while improving soil fertility. Similarly, in pre-Columbian America, indigenous peoples practiced a form of silvopasture known as the *"chinampa"* system, which involved the cultivation of crops on raised beds within shallow lake areas, with trees planted alongside to provide fodder for animals.

As human populations grew and agricultural practices expanded, silvopasture systems continued to evolve. In medieval Europe, the practice of *"wood-pasture"* emerged, where livestock grazed in forests and woodlands, benefiting from the diverse forage and shelter provided by the trees. This system was particularly prevalent in England, where it played a crucial role in shaping the landscape and supporting rural communities.

The geographical and climatic conditions of different regions also influenced the development of silvopasture systems. In the arid and semi-arid regions of Africa and Asia, for example, pastoralists developed systems that integrated trees and shrubs with grazing lands to provide shade, shelter, and fodder for their animals. These systems, known as *"parklands,"* allowed communities to maintain their livestock even in harsh environments with limited resources.

In more recent times, the modernization of agriculture and the rise of industrial farming practices led to a decline in traditional silvopasture systems. However, in the latter part of the *20th century*, there was a resurgence of interest in these systems to promote sustainable agriculture and address environmental challenges. This renewed interest has led to the development of innovative silvopasture techniques, such as planting fast-growing tree species for fodder production, using rotational grazing to improve soil health, and integrating agroforestry practices with modern livestock management.

Today, silvopasture systems are recognized for their potential to **enhance agricultural productivity**, **improve animal welfare**, and contribute to **climate change mitigation**. By sequestering carbon in the soil and trees, these systems can help to offset greenhouse gas emissions from livestock production. Furthermore, silvopasture can play a critical role in promoting biodiversity and preserving ecosystems, as the integration of trees and shrubs with grazing lands creates habitats for a wide range of flora and fauna.

In conclusion, the development of silvopasture systems has been shaped by a complex interplay of historical, cultural, and environmental factors. As we look to the future, it is clear that these systems will continue to evolve and adapt, playing a vital role in pursuing sustainable agriculture and mitigating climate change.

Modern Agroforestry: Innovations and Challenges

In recent years, agroforestry has experienced a resurgence as a viable and sustainable land management practice. This renewed interest is driven by the need for innovative solutions to address pressing global challenges, such as food security, climate change, and environmental degradation. In this section, we will explore the modern innovations in agroforestry and the challenges faced in implementing these practices.

One of the most significant innovations in modern agroforestry is the development of new, highly productive **tree species and cultivars**. These improved varieties have been bred for specific traits, such as rapid growth, high biomass production, and resistance to pests and diseases. This has allowed for the integration of trees into agricultural systems with minimal impact on crop yields while providing additional benefits such as timber, fuelwood, and non-timber forest products.

Another innovation in agroforestry is using advanced spatial planning and design tools. **Geographic Information Systems (GIS)** and **remote sensing** technologies have enabled researchers and practitioners to map and analyze complex agroforestry systems. This allows a better understanding of the interactions between trees, crops, and livestock. This has led to the development of more efficient and sustainable land-use strategies, such as optimizing tree placement to maximize benefits and minimize competition with crops.

In addition to these technological advancements, modern agroforestry has also benefited from a growing body of scientific research. This research has helped to quantify the ecological, economic, and social benefits of agroforestry practices, providing evidence-based support for their adoption and implementation. For example, studies have shown that agroforestry systems can **increase biodiversity, improve soil health**, and **enhance ecosystem services** such as carbon sequestration and water regulation.

Despite these innovations, several challenges still need to be addressed to fully realize the potential of agroforestry. One of the primary challenges is the lack of awareness and understanding of agroforestry practices among farmers, policymakers, and the general public. This can lead to skepticism and resistance to adopting new land management practices, even when they have been proven to be beneficial.

Another challenge is the need for supportive policies and incentives to encourage the adoption of agroforestry practices. In many countries, agricultural policies and subsidies are still heavily focused on monoculture systems, which can discourage farmers from diversifying their land use. To overcome this barrier, governments and international organizations must develop and implement policies that recognize and reward the benefits of agroforestry systems.

Lastly, there is a need for continued research and innovation to address the specific challenges faced by different agroforestry systems in various regions and contexts. This includes developing new tree species and cultivars that are adapted to changing climate conditions, as well as refining management practices to optimize productivity and sustainability.

In conclusion, modern agroforestry has come a long way from its ancient roots, with significant innovations in tree breeding, spatial planning, and scientific research. However, there are still challenges to be addressed in order to fully harness the potential of agroforestry as a sustainable land management practice. By overcoming these challenges and continuing to innovate, agroforestry can play a crucial role in addressing global food security, climate change, and environmental degradation.

The Role of Agroforestry in Sustainable Agriculture and Climate Change Mitigation

As the world grapples with the pressing issues of climate change and food security, agroforestry has emerged as a vital solution that addresses both concerns. By integrating trees into agricultural landscapes, agroforestry promotes sustainable agriculture and plays a significant role in mitigating the effects of climate change. This section delves into the various ways in which agroforestry contributes to these global challenges.

First and foremost, agroforestry enhances the productivity and resilience of agricultural systems. The presence of trees in farmlands provides numerous benefits, such as improving **soil fertility, reducing soil erosion**, and **regulating microclimates.** These factors contribute to increased crop yields and a more stable food supply, which is crucial in the face of a growing global population. Furthermore, the diversification of income sources through agroforestry, such as timber and non-timber forest products, strengthens the economic resilience of farmers and rural communities.

In terms of climate change mitigation, agroforestry systems have a remarkable capacity to sequester carbon. Trees absorb and store atmospheric carbon dioxide, effectively reducing greenhouse gas emissions. Studies have shown that agroforestry systems can sequester up to four times more carbon than conventional agricultural systems. This carbon sequestration potential makes agroforestry a critical tool in the global effort to combat climate change.

Additionally, agroforestry systems can help to conserve and enhance biodiversity. By creating diverse and complex habitats, agroforestry supports a wide range of flora and fauna, including pollinators, natural predators, and other beneficial organisms. This increased biodiversity not only contributes to the overall health of ecosystems but also provides essential ecosystem services, such as **pest control** and **pollination**, which are vital for sustainable agriculture.

Agroforestry also plays a role in climate change adaptation. The integration of trees into agricultural systems can help buffer against extreme weather events, such as droughts, floods, and storms. Trees can act as windbreaks, protecting crops from damaging winds, and their root systems can help to stabilize soils, reducing the risk of landslides and erosion. Moreover, the shade provided by trees can help to moderate temperatures, reducing heat stress on crops and livestock.

Lastly, agroforestry can contribute to the sustainable management of water resources. For example, trees can improve water infiltration and retention in soils, reducing surface runoff and the risk of flooding. They can also help to maintain water quality by filtering pollutants and preventing sedimentation in water bodies. These benefits are particularly important in increasing water scarcity and the need for efficient water use in agriculture.

In conclusion, agroforestry is crucial in promoting sustainable agriculture and mitigating the effects of climate change. By integrating trees into agricultural systems, agroforestry offers a multitude of environmental, social, and economic benefits that contribute to a more resilient and sustainable future. As we look ahead, the continued adoption and innovation of agroforestry and silvopasture practices will be essential in addressing the global challenges of food security and climate change.

The Future of Agroforestry and Silvopasture

As we reach the end of our journey through the history and evolution of agroforestry and silvopasture, it is essential to look ahead and envision the future of these sustainable land management practices. In a world grappling with the consequences of climate change, deforestation, and food insecurity, the importance of agroforestry and silvopasture cannot be overstated. These practices offer a promising path toward a more sustainable and resilient future for agriculture, ecosystems, and global communities.

The future of agroforestry and silvopasture will likely be shaped by several critical factors, including technological advancements, policy support, and increased awareness and adoption by farmers and landowners. As technology continues to advance, we can expect innovations to emerge that will further enhance the efficiency and effectiveness of agroforestry and silvopasture systems. For instance, precision agriculture tools, such as drones and satellite imagery, can help farmers monitor and manage their agroforestry systems more effectively, while advancements in plant breeding and genetics can lead to the development of more productive and resilient tree and crop species.

Policy support will also play a crucial role in the future of agroforestry and silvopasture. Governments and international organizations must recognize the value of these practices in promoting sustainable agriculture, conserving biodiversity, and mitigating climate change. By providing financial incentives, technical assistance, and favorable regulatory frameworks, policymakers can encourage the widespread adoption of agroforestry and silvopasture systems. Additionally, integrating agroforestry and silvopasture into national and international climate change strategies can help countries meet their greenhouse gas emissions reduction targets while enhancing food security and rural livelihoods.

Increased awareness and adoption of agroforestry and silvopasture practices by farmers and landowners is another critical factor for the future of these systems. Education and outreach efforts must be scaled up to ensure that farmers are aware of the benefits of agroforestry and silvopasture and have access to the knowledge and resources needed to implement these practices successfully. This can be achieved through farmer-to-farmer knowledge exchange, capacity-building programs, and the development of demonstration sites showcasing the benefits of agroforestry and silvopasture systems.

In conclusion, the future of agroforestry and silvopasture is bright, with immense potential for transforming agriculture and land management practices worldwide. By embracing these sustainable systems, we can work towards a future where agriculture and forestry coexist harmoniously, providing food, income, and ecosystem services while also mitigating the impacts of climate change. The task ahead is not an easy one, but with continued innovation, policy support, and widespread adoption, agroforestry and silvopasture can play a pivotal role in shaping a more sustainable and resilient world for generations to come.

Chapter Summary

- Agroforestry, the integration of trees and shrubs into agricultural systems, has been practiced by various cultures across the globe since the dawn of agriculture, providing numerous benefits such as improved soil fertility, enhanced biodiversity, and increased resilience against climate change.
- Ancient civilizations, including the indigenous cultures of the Amazon Basin, ancient African societies, the Mayans, and traditional rice paddy systems in Asia, all utilized agroforestry techniques to sustain their communities and foster environmental stewardship.
- Silvopasture, a practice that combines forestry and the grazing of domesticated animals, has a rich and diverse history, with roots in ancient civilizations and evolving over time to suit different environments and cultures.
- Modern agroforestry has experienced a resurgence due to the need for innovative solutions to address pressing global challenges, such as food security, climate change, and environmental degradation.
- Innovations in modern agroforestry include the development of new, highly productive tree species and cultivars, advanced spatial planning and design tools, and a growing body of

scientific research that quantifies the ecological, economic, and social benefits of agroforestry practices.

- Challenges faced in implementing agroforestry practices include a lack of awareness and understanding among farmers, policymakers, and the general public, the need for supportive policies and incentives, and the need for continued research and innovation to address specific challenges in different regions and contexts.
- Agroforestry is crucial in promoting sustainable agriculture and mitigating the effects of climate change by enhancing productivity and resilience, sequestering carbon, conserving and enhancing biodiversity, and contributing to climate change adaptation and sustainable water resource management.
- The future of agroforestry and silvopasture will be shaped by technological advancements, policy support, and increased awareness and adoption by farmers and landowners, with the potential to transform agriculture and land management practices worldwide and contribute to a more sustainable and resilient future.

2

PRINCIPLES AND CONCEPTS OF AGROFORESTRY

An image that showcases the symbiotic relationship between trees, crops, and livestock in a thriving agroforestry system, highlighting the interdependence and harmony between these elements.

Agroforestry, an age-old practice that has been refined over generations, is a sustainable land management system that integrates trees, crops, and livestock on the same piece of land. This harmonious coexistence of diverse species optimizes the use of natural resources and offers many ecological, social, and economic benefits. In this chapter, we will delve into the principles and concepts that underpin agroforestry, providing a solid foundation for understanding the intricacies of this fascinating approach to land management.

At the heart of agroforestry lies the idea of **working with nature** rather than against it. By mimicking the structure and function of natural ecosystems, agroforestry systems are designed to enhance productivity, resilience, and sustainability. This is achieved through carefully selecting and arranging tree, crop, and livestock components, which are chosen based on their complementary traits and interactions. The result is a dynamic, self-regulating system that provides multiple benefits while minimizing the need for external inputs and reducing the risk of environmental degradation.

One of the key principles of agroforestry is **diversity**. By incorporating a wide range of plant and animal species, agroforestry systems can capitalize on the unique strengths and characteristics of each component while also buffering against potential threats such as pests, diseases, and climate variability. This diversity also promotes beneficial interactions among species, such as nutrient cycling, pollination, and natural pest control, which contribute to the overall health and productivity of the system.

Another important concept in agroforestry is **spatial and temporal integration**. This refers to the strategic arrangement of trees, crops, and livestock in space and time to maximize their complementary interactions and minimize competition for resources such as light, water, and nutrients. For example, trees can be planted in rows or clusters to provide shade and wind protection for crops while allowing enough sunlight to penetrate the canopy for optimal growth. Similarly, livestock can be rotated through different parts of the agroforestry system to help control weeds, fertilize the soil, and reduce the buildup of pests and diseases.

In addition to these ecological principles, agroforestry encompasses various social and economic considerations. For instance, agroforestry systems must be designed to meet the needs and preferences of local communities, taking into account factors such as land tenure, labor availability, and market opportunities. By providing a diverse array of products and services, agroforestry can help to enhance food security, generate income, and improve the livelihoods of rural households while also contributing to broader goals such as climate change mitigation and biodiversity conservation.

In the following sections, we will explore the ecological foundation of agroforestry systems, the key components and design considerations involved, and the role of agroforestry in sustainable land management. We will also examine the socioeconomic benefits and challenges associated with agroforestry adoption and discuss the prospects of agroforestry and silvopasture in the face of global environmental change.

The Ecological Foundation of Agroforestry Systems

The ecological foundation of agroforestry systems is rooted in the harmonious integration of trees, crops, and livestock, which work together to create a sustainable and productive landscape. By mimicking natural ecosystems, agroforestry systems promote biodiversity, enhance soil health, and contribute to climate change mitigation. In this section, we will delve into the ecological principles that underpin agroforestry systems and explore how these principles contribute to the overall sustainability and resilience of the landscape.

One of the core ecological principles of agroforestry is the concept of **diversity**. In nature, diverse ecosystems are more resilient and productive than monocultures. Agroforestry systems capitalize on this principle by integrating a variety of plant and animal species, each with its unique roles and functions. This diversity not only enhances the system's overall productivity but also reduces the risk of pests and diseases, as well as the reliance on external inputs such as fertilizers and pesticides.

Another fundamental ecological principle in agroforestry is the **efficient use of resources**. By integrating trees, crops, and livestock, agroforestry systems can optimize the use of available resources such as sunlight, water, and nutrients. For example, trees can provide shade and shelter for crops and livestock, reducing the need for irrigation and protecting against extreme weather events. In addition, trees and crops can be strategically planted to improve nutrient cycling and reduce soil erosion, leading to healthier and more fertile soils.

The ecological foundation of agroforestry systems also includes the principle of **mutualism** and **symbiosis.** In these systems, different species interact in ways that benefit one another, creating a more resilient and productive landscape. For instance, nitrogen-fixing trees can improve soil fertility by converting atmospheric nitrogen into a form that crops can utilize. Similarly, livestock can contribute to nutrient cycling and soil fertility by grazing on crop residues and depositing manure.

Agroforestry systems are also crucial in **carbon sequestration** and **climate change mitigation.** Trees and other perennial plants in these systems absorb and store atmospheric carbon dioxide, helping to reduce greenhouse gas emissions. Moreover, agroforestry practices can enhance the overall resilience of the landscape to climate change by promoting biodiversity, improving water and nutrient cycling, and reducing the risk of soil erosion.

In summary, the ecological foundation of agroforestry systems is built upon the principles of diversity, resource efficiency, mutualism, and climate change mitigation. By integrating trees, crops, and livestock in a manner that mimics natural ecosystems, agroforestry systems contribute to the overall sustainability and resilience of the landscape. In the following sections, we will explore the key components and design considerations of agroforestry systems and their role in sustainable land management and socioeconomic development.

Key Components and Design Considerations in Agroforestry

The success of an agroforestry system depends on the careful selection and integration of its key components, which include trees, crops, and livestock. These components must be chosen and arranged to optimize the system's ecological and socioeconomic benefits. This section will discuss the essential elements of agroforestry and the design considerations that must be considered when planning and implementing an agroforestry system.

Trees: Trees are the backbone of any agroforestry system, providing many ecological and economic benefits. When selecting tree species for an agroforestry system, it is crucial to consider factors such as growth rate, canopy structure, root system, and the ability to fix nitrogen. Additionally, the choice of tree species should be based on their compatibility with the local climate, soil conditions, and the specific needs of the crops and livestock in the system.

Crops: Crops are an integral part of agroforestry systems, providing food, fodder, and income for farmers. The selection of crops should be based on their adaptability to the local climate and soil conditions and their compatibility with the chosen tree species. When selecting crops for an agroforestry system, it is essential to consider factors such as light requirements, water needs, and nutrient demands.

Livestock: Incorporating livestock into an agroforestry system can enhance its productivity and sustainability by providing additional income sources, improving nutrient cycling, and promoting natural pest control. The choice of livestock species should be based on their compatibility with the tree and crop components and their ability to thrive in the local environment.

Design Considerations

When designing an agroforestry system, several factors must be considered to ensure its success. These include:

- **Spatial arrangement:** The layout of trees, crops, and livestock should be planned to optimize the use of space, light, and resources while minimizing competition between components.
- **Integration of components:** The interactions between trees, crops, and livestock should be carefully considered to promote synergies and minimize negative impacts. For example,

selecting trees that provide shade and shelter for crops and livestock can enhance their productivity and well-being.

- **Management practices:** The choice of management practices, such as pruning, fertilization, and pest control, should be tailored to the specific needs of the agroforestry system and its components.
- **Market considerations:** The choice of tree, crop, and livestock species should be based on their market potential and the needs of the local community, ensuring that the system provides a reliable source of income for farmers.

In conclusion, the key components and design considerations in agroforestry systems are crucial for their success and sustainability. By carefully selecting and integrating trees, crops, and livestock and considering factors such as spatial arrangement, management practices, and market potential, agroforestry systems can provide various ecological and socioeconomic benefits for farmers and the environment.

The Role of Agroforestry in Sustainable Land Management

In today's world, sustainable land management has become crucial to preserving our environment and ensuring food security for future generations. Agroforestry, an integrative approach that combines agricultural and forestry practices, plays a vital role in achieving this goal. This section will explore how agroforestry contributes to sustainable land management through its various ecological, economic, and social benefits.

One of the primary ways agroforestry promotes sustainable land management is through its ability to **enhance soil health.** Integrating trees and shrubs in agricultural systems helps improve soil fertility by fixing atmospheric nitrogen, recycling nutrients, and increasing organic matter content. This, in turn, leads to improved soil structure, water retention, and reduced erosion. Moreover, the deep-rooting nature of trees helps to prevent soil compaction and maintain a healthy soil profile, which is essential for the long-term productivity of the land.

Agroforestry systems also contribute to sustainable land management by promoting **biodiversity**. Integrating various tree species and crops creates a diverse and dynamic ecosystem that supports a wide range of flora and fauna. This increased biodiversity enhances the system's resilience against pests and diseases and provides essential habitat for pollinators and other beneficial organisms. Furthermore, agroforestry can serve as a valuable tool for landscape connectivity, creating corridors for wildlife movement and promoting the conservation of native species.

Another significant aspect of agroforestry's role in sustainable land management is its potential for **climate change mitigation** and adaptation. Trees in agroforestry systems act as carbon sinks, sequestering atmospheric carbon dioxide and reducing greenhouse gas emissions. Additionally, the increased tree cover provides shade and microclimatic benefits, helping to regulate temperature and humidity within the system. This can be particularly beneficial in the face of increasing temperatures and erratic rainfall patterns associated with climate change. Agroforestry systems also contribute to climate change adaptation by enhancing the resilience of agricultural landscapes to extreme weather events, such as floods and droughts, through improved soil health and water management.

From an economic perspective, agroforestry can contribute to sustainable land management by diversifying farm income and reducing the risks associated with monoculture systems. The integration of trees and crops allows farmers to tap into multiple markets and income streams, providing a safety net during crop failure or fluctuating market prices. Furthermore, agroforestry products, such as timber, fruits, and nuts, often fetch higher market prices than traditional agricultural commodities, contributing to increased farm profitability and long-term viability.

Lastly, agroforestry plays a crucial role in sustainable land management by addressing various social and cultural aspects of rural communities. The adoption of agroforestry practices can lead to improved **food security and nutrition**, as well as increased access to fuelwood, fodder, and other essential resources. Moreover, agroforestry systems can contribute to preserving traditional knowledge and cultural practices, fostering a sense of pride and identity among local communities.

In conclusion, agroforestry is a powerful tool for sustainable land management, offering a range of ecological, economic, and social benefits. By integrating trees and crops in a unified and dynamic system, agroforestry can help to preserve our environment, enhance food security, and promote the well-being of rural communities. As we look towards the future, the continued adoption and development of agroforestry and silvopasture practices will be essential in addressing the complex challenges of climate change, biodiversity loss, and global food security.

Socioeconomic Benefits and Challenges of Agroforestry Adoption

As an innovative and sustainable land management practice, agroforestry offers myriad socioeconomic benefits to farmers, communities, and nations. However, it also presents certain challenges that must be addressed for successful adoption and implementation. This section will explore the various socioeconomic benefits and challenges of agroforestry adoption.

One of the most significant benefits of agroforestry is its potential to **diversify** and **increase farmers' income.** By integrating trees and crops, farmers can generate multiple sources of revenue from the same piece of land. This diversification provides financial stability and reduces the risks associated with monoculture farming, such as crop failure due to pests, diseases, or adverse weather conditions. Additionally, agroforestry systems can provide valuable non-timber forest products (NTFPs), such as fruits, nuts, medicinal plants, and fodder, which farming households can sell or consume.

Agroforestry also contributes to **food security** by enhancing the productivity and resilience of agricultural systems. The integration of trees in farming systems can improve soil fertility, reduce soil erosion, and provide shade and windbreaks, which in turn can increase crop yields. Moreover, agroforestry can help maintain and enhance biodiversity, which is crucial for the long-term sustainability of agriculture and food production.

Another important socioeconomic benefit of agroforestry is its potential to create **employment opportunities**, particularly in rural areas. The labor-intensive nature of agroforestry practices, such as planting, pruning, and harvesting, can generate jobs for local communities. Additionally, the processing and marketing of agroforestry products can stimulate the growth of small and medium enterprises, further contributing to rural development and poverty alleviation.

Despite these numerous benefits, the adoption of agroforestry also presents some challenges. One of the main challenges is the need for more knowledge and technical expertise among farmers and extension agents. The successful implementation of agroforestry systems requires a deep understanding of the ecological and agronomic principles involved and the ability to select appropriate tree species and management practices. This knowledge gap can be addressed through capacity building, training, and the dissemination of agroforestry technologies and best practices.

Another challenge is the relatively long-term nature of agroforestry investments. Trees usually take several years to mature and provide significant economic returns, which may discourage farmers from adopting agroforestry practices. To overcome this barrier, we must promote short-term benefits, such as the provision of fodder, firewood, or NTFPs, and develop financial mechanisms that support long-term investments in agroforestry.

Lastly, the lack of supportive policies and institutional frameworks can hinder the widespread adoption of agroforestry. Therefore, governments and policymakers need to recognize the value of agroforestry in addressing various socioeconomic and environmental challenges and create enabling conditions for its promotion and implementation. This may include the development of agroforestry-specific policies, the integration of agroforestry into existing agricultural and forestry policies, and the provision of incentives for farmers to adopt agroforestry practices.

In conclusion, agroforestry offers a multitude of socioeconomic benefits, including income diversification, food security, and rural development. However, to fully harness its potential, it is crucial to address the challenges related to knowledge, long-term investments, and policy support. By doing so, agroforestry can play a vital role in promoting sustainable land management and improving the livelihoods of millions of people around the world.

The Future of Agroforestry and Silvopasture

As explored throughout this chapter, agroforestry and silvopasture offer many ecological, economic, and social benefits that can contribute to a more sustainable and resilient future for agriculture and land management. By integrating trees, crops, and livestock, these systems can transform how we produce food, fiber, and fuel while providing essential ecosystem services and fostering biodiversity.

The future of agroforestry and silvopasture is bright. More and more farmers, landowners, and policymakers recognize the value of these practices in addressing pressing global challenges such as climate change, food security, and rural development. As we move forward, it is essential to continue advancing our understanding of these systems and refining their design and management to maximize their potential benefits.

One crucial aspect of this endeavor is the ongoing development and dissemination of agroforestry knowledge and best practices. This includes not only scientific research and technical guidance but also the sharing of traditional and indigenous knowledge that has been honed over generations of working with the land. By fostering a global community of agroforestry practitioners and researchers, we can ensure that these systems continue to evolve and adapt to the changing needs of our world.

Another critical factor in the future of agroforestry and silvopasture is the development of supportive policies and incentives at local, national, and international levels. This includes measures such as financial support for agroforestry adoption, the integration of agroforestry into agricultural and environmental policies, and the recognition of agroforestry's contributions to climate change mitigation and adaptation efforts. By creating an enabling environment for agroforestry, we can accelerate its adoption and maximize its impact.

Finally, the future of agroforestry and silvopasture will be shaped by the choices and actions of individual farmers, landowners, and communities. By embracing these systems and incorporating them into their land management strategies, these stakeholders can help to drive a shift towards more sustainable and resilient agricultural landscapes. This, in turn, will contribute to improved livelihoods, enhanced food security, and a healthier planet for all.

In conclusion, agroforestry and silvopasture represent a promising pathway toward a more sustainable and resilient future for agriculture and land management. By integrating trees, crops, and livestock, these systems can help address some of our most pressing challenges while providing a wide range of ecological, economic, and social benefits. As we continue to advance our understanding and practice of agroforestry, we can look forward to a future in which these systems play an increasingly important role in shaping the health and well-being of our planet and its inhabitants.

Chapter Summary

- Agroforestry is a sustainable land management system that integrates trees, crops, and livestock on the same piece of land, offering ecological, social, and economic benefits.
- The principles of agroforestry include diversity, spatial and temporal integration, and working with nature to mimic natural ecosystems, enhancing productivity, resilience, and sustainability.
- Agroforestry systems promote biodiversity, enhance soil health, and contribute to climate change mitigation by sequestering carbon dioxide and improving landscape resilience.
- Key components of agroforestry systems include trees, crops, and livestock, which must be carefully selected and integrated based on their complementary traits and interactions.
- Agroforestry is vital in sustainable land management by improving soil fertility, promoting biodiversity, and mitigating climate change impacts.

- Socioeconomic benefits of agroforestry include income diversification, food security, rural development, and employment opportunities.
- Challenges to agroforestry adoption include a need for knowledge and technical expertise, long-term investment nature, and insufficient supportive policies and institutional frameworks.
- The future of agroforestry and silvopasture is promising, with continued development and dissemination of knowledge, supportive policies, and individual actions driving a shift towards more sustainable and resilient agricultural landscapes.

3

SILVOPASTURE: INTEGRATING LIVESTOCK, TREES, AND FORAGE

A harmonious silvopasture landscape, where livestock, trees, and forage coexist to create a sustainable and productive ecosystem.

In today's rapidly changing world, the need for sustainable agricultural practices has never been more critical. As the global population continues to grow and climate change threatens the stability of our ecosystems, innovative solutions are required to ensure food security while minimizing environmental impacts. One such solution is silvopasture, a sustainable approach to agriculture that integrates livestock, trees, and forage into a single, harmonious system.

Silvopasture is not a new concept; in fact, it has been practiced for centuries by farmers and indigenous communities around the world. However, it is only recently that this ancient practice has gained renewed attention as a viable and sustainable alternative to conventional agriculture. By combining the best of both worlds – the productive capacity of livestock farming and the ecological benefits of forestry – silvopasture offers a unique and promising solution to some of the most pressing challenges facing modern agriculture.

At its core, silvopasture is about **creating a diverse and resilient agricultural landscape** that supports multiple functions and provides a wide range of benefits. This is achieved by carefully selecting and managing the right combination of tree species, forage plants, and livestock to create a productive and self-sustaining system. When done correctly, silvopasture can improve soil health, increase biodiversity, enhance animal welfare, and even boost farm profitability.

In this chapter, we will explore the many benefits of silvopasture, from its environmental advantages to its economic and social contributions. We will delve into the key components of a successful silvopasture system and provide practical guidance on designing and implementing this innovative approach to agriculture. Through real-world case studies, we will demonstrate the transformative potential of silvopasture in action, showcasing its ability to create more resilient and sustainable agricultural systems.

As we embark on this journey into the world of silvopasture, we hope that you will come away with a deeper understanding of this sustainable approach to agriculture and be inspired to consider its potential in your farming endeavors. The future of agriculture depends on our ability to adapt and innovate, and silvopasture offers a promising path forward for a more sustainable and resilient food system.

Benefits of Silvopasture: Environmental, Economic, and Social Advantages

Silvopasture, a sustainable agricultural practice that combines livestock, trees, and forage, offers a myriad of benefits that extend beyond the farm. By integrating these elements, silvopasture systems create a harmonious balance that benefits the environment, the economy, and society as a whole. This section will delve into the numerous advantages of adopting silvopasture practices.

Environmental Benefits

One of the most significant benefits of silvopasture is its positive impact on the environment. By incorporating trees into livestock systems, silvopasture helps combat climate change through carbon sequestration. Trees absorb carbon dioxide from the atmosphere and store it in their biomass, effectively reducing greenhouse gas emissions.

Additionally, silvopasture systems promote biodiversity by providing a diverse habitat for various flora and fauna. Trees and shrubs create a more complex ecosystem, attracting pollinators, birds, and other wildlife. This increased biodiversity contributes to more robust and resilient ecosystems, which can better withstand the impacts of climate change and other environmental stressors.

Silvopasture also helps to conserve water resources. Tree roots help to retain water in the soil, reducing runoff and erosion. This conserves water and improves water quality by filtering out pollutants and sediment.

Economic Benefits

Silvopasture systems offer numerous economic advantages for farmers and landowners. By diversifying farm income, silvopasture can provide a more stable financial foundation. Livestock, timber, and forage production can all generate income, reducing reliance on a single commodity.

Furthermore, the integration of trees and livestock can lead to increased productivity. Trees provide shade and shelter for livestock, reducing heat stress and improving animal welfare. This, in turn, can lead to higher weight gains and improved reproductive performance, ultimately increasing the farm's overall profitability.

Silvopasture can also reduce input costs for farmers. The presence of trees and shrubs can reduce the need for supplemental feed, as they provide additional forage for livestock. Moreover, trees can act as natural windbreaks, reducing the need for expensive infrastructure to protect livestock from harsh weather conditions.

Social Benefits

Silvopasture systems contribute to the social well-being of rural communities by promoting food security and preserving cultural heritage. In addition, by diversifying agricultural production, silvopasture can help to ensure a stable food supply, even in the face of climate change and other challenges.

Additionally, silvopasture practices often align with traditional land management techniques, allowing farmers to maintain their **cultural identity** while embracing sustainable agriculture. This can foster a sense of pride and connection to the land, strengthening rural communities.

In conclusion, silvopasture offers many environmental, economic, and social benefits, making it an attractive option for farmers and landowners seeking sustainable agricultural practices. By integrating livestock, trees, and forage, silvopasture systems create a harmonious balance that benefits the farm, the broader environment, and society.

Key Components of a Successful Silvopasture System

Silvopasture is a dynamic and innovative approach to agriculture that combines the best of both worlds: the productive capacity of livestock farming and the ecological benefits of tree cultivation. By integrating these elements, silvopasture systems can create a sustainable and resilient agricultural landscape that provides multiple benefits for farmers, livestock, and the environment. In this section, we will discuss the key components that contribute to the success of a silvopasture system, ensuring its long-term viability and effectiveness.

Careful Selection of Tree Species: The choice of tree species is a crucial factor in the success of a silvopasture system. Trees should be selected based on their compatibility with the local climate, soil conditions, and the livestock's specific needs. Ideally, the chosen tree species should provide shade, shelter, and forage for the animals while contributing to soil fertility and carbon sequestration.

Strategic Tree Planting and Spacing: The arrangement and density of trees in a silvopasture system can significantly impact its overall productivity and functionality. Trees should be planted in a way that maximizes the available sunlight for both the trees and the understory forage while also providing adequate space for livestock to graze and move freely. Proper tree spacing can also help prevent soil compaction and erosion and promote healthy root development and nutrient cycling.

Diverse and Nutritious Forage: A successful silvopasture system relies on a diverse and nutritious forage base to support the health and productivity of the livestock. This can include a mix of grasses, legumes, and other herbaceous plants that provide a balanced diet for the animals and a habitat for beneficial insects and other wildlife. Regular monitoring and management of the forage are essential to ensure its quality and availability throughout the grazing season.

Adaptive Livestock Management: Integrating livestock into a silvopasture system requires careful planning and adaptive management to ensure the animals' well-being and the system's long-term sustainability. This includes selecting appropriate livestock species and breeds, implementing rotational grazing practices, and monitoring animal health and performance. By adapting management practices to the specific needs of the silvopasture system, farmers can optimize livestock productivity while minimizing negative impacts on the environment.

Ongoing Monitoring and Evaluation: A successful silvopasture system is one that evolves and adapts over time, responding to changing conditions and the needs of the farm. Regular monitoring and evaluation of the system's performance can help identify areas for improvement and inform future management decisions. This may involve tracking tree growth and health, forage quality and availability, livestock performance, and other indicators of system health and productivity.

In conclusion, the key components of a successful silvopasture system include careful selection of tree species, strategic tree planting and spacing, diverse and nutritious forage, adaptive livestock management, and ongoing monitoring and evaluation. By incorporating these elements into the design and management of a silvopasture system, farmers can create a resilient and sustainable agricultural landscape that benefits both people and the environment.

Designing and Implementing Silvopasture: Best Practices and Strategies

Silvopasture, a sustainable approach to agriculture, requires careful planning and implementation to ensure its success. By integrating livestock, trees, and forage, farmers can create a harmonious and productive ecosystem that benefits the environment, the economy, and society. This section will explore best practices and strategies for designing and implementing a successful silvopasture system.

Assessing the Site: Before embarking on a silvopasture project, conducting a thorough site assessment is crucial. This includes evaluating the soil type, topography, climate, and existing vegetation. Understanding these factors will help you determine the most suitable tree species, livestock, and forage crops for your system.

Choosing the Right Tree Species: The selection of tree species is a critical factor in the success of a silvopasture system. Consider native species that are well-adapted to the local climate and soil conditions. Additionally, choose trees that provide multiple benefits, such as shade, timber, and fodder for livestock.

Integrating Livestock: The choice of livestock should be based on the available resources, market demand, and the compatibility of the animals with the chosen tree species. It is essential to ensure that the livestock's grazing habits do not negatively impact the growth and development of the trees.

Forage Management: A successful silvopasture system requires careful management of the forage component. This includes selecting appropriate forage species, implementing rotational grazing practices, and monitoring the health and productivity of the forage crops.

Tree Protection and Maintenance: Young trees in a silvopasture system are vulnerable to damage from livestock. Employ protective measures, such as tree guards or temporary fencing, to prevent browsing and trampling. Regular pruning and thinning of the trees will also help to maintain their health and productivity.

Monitoring and Evaluation: Regular monitoring of the silvopasture system is essential to ensure its success. Keep track of tree growth, forage production, and livestock health to identify any issues that may arise. Adjust management practices as needed to maintain the system's productivity and sustainability.

Education and Training: Silvopasture is a complex and dynamic system that requires specialized knowledge and skills. Invest in education and training for yourself and your team to ensure that you are equipped to manage the system effectively.

By following these best practices and strategies, you can design and implement a successful silvopasture system that provides various environmental, economic, and social benefits. As more farmers adopt this sustainable approach to agriculture, we can look forward to a future where our agricultural landscapes are more resilient, diverse, and productive.

Case Studies: Real-World Examples of Silvopasture in Action

Silvopasture systems have been successfully implemented in various regions around the world, showcasing the adaptability and effectiveness of this sustainable agricultural practice. In this section, we will explore several real-world examples of silvopasture in action, highlighting the unique challenges and successes experienced by farmers and land managers who have embraced this innovative approach to agriculture.

Case Study 1: Pine Forest Silvopasture in the Southeastern United States

In the southeastern United States, a growing number of farmers are integrating livestock, trees, and forage crops in a silvopasture system based on pine forests. This practice involves thinning existing pine stands to allow for the growth of forage crops, such as grasses and legumes, which provide nutritious feed for grazing animals like cattle and sheep. The trees offer shade and shelter for the livestock, while their roots help to prevent soil erosion and improve nutrient cycling. Farmers in this region have reported increased livestock productivity, improved timber quality, and enhanced wildlife habitat and carbon sequestration benefits.

Case Study 2: Coffee Agroforestry in Central America

In Central America, coffee farmers have been integrating shade trees and livestock into their coffee plantations to create a more resilient and sustainable agroecosystem. By incorporating trees and shrubs that provide shade, farmers can reduce the need for chemical inputs, protect their crops from pests and diseases, and improve soil fertility. Livestock, such as chickens and pigs, can be raised alongside the coffee plants, providing additional income and contributing to the farm's overall productivity. This silvopasture system has been shown to increase coffee yields, improve the quality of the beans, and enhance biodiversity.

Case Study 3: Temperate Silvopasture in the United Kingdom

In the United Kingdom, some farmers are experimenting with silvopasture systems in temperate climates, combining livestock grazing with the cultivation of fruit and nut trees. This approach allows farmers to diversify their income streams and reduce reliance on monoculture cropping systems. By integrating trees and livestock, farmers can improve soil fertility, reduce erosion, and create habitats for beneficial insects and wildlife. The trees also provide valuable timber and non-timber forest products, such as fruits, nuts, and firewood, which can be sold or used on the farm.

Case Study 4: Tropical Silvopasture in Brazil

In Brazil, silvopasture systems have been used to restore degraded pasturelands and improve the productivity of cattle ranching operations. By planting native tree species alongside forage grasses, ranchers can create a more diverse and resilient landscape that supports both livestock and wildlife. The trees provide shade, reducing heat stress for the animals, and their deep roots help to improve soil structure and water infiltration. This approach has been shown to increase livestock weight gain and overall productivity while providing essential ecosystem services, such as carbon sequestration and habitat for native species.

These case studies demonstrate the versatility and effectiveness of silvopasture systems in a variety of contexts and climates. By integrating livestock, trees, and forage crops, farmers and land managers can create more sustainable, productive, and resilient agricultural systems that benefit both people and the environment. As the world faces increasing challenges related to climate change, resource scarcity, and food security, silvopasture offers a promising solution for the future of agriculture.

The Future of Silvopasture and Its Role in Sustainable Agriculture

In conclusion, silvopasture is a promising and innovative approach to sustainable agriculture, offering many environmental, economic, and social benefits. By integrating livestock, trees, and forage into a single, harmonious system, silvopasture can revolutionize how we approach farming and land management. As we face increasing challenges from climate change, population growth, and resource scarcity, the need for sustainable agricultural practices has never been more critical.

The future of silvopasture is bright; more farmers and landowners recognize its advantages and adopt this practice on their lands. Through continued research, education, and collaboration, we can refine and optimize silvopasture systems to maximize their benefits and minimize potential drawbacks. Furthermore, by **sharing knowledge and experiences**, we can help ensure that silvopasture becomes an integral part of sustainable agriculture worldwide.

As we have explored in this chapter, the key components of a successful silvopasture system include careful planning, design, implementation, ongoing management, and monitoring. By adhering to best practices and strategies, farmers can create thriving silvopasture systems that benefit their livestock, trees, forage, the environment, and their bottom line.

The case studies presented in this chapter demonstrate the versatility and adaptability of silvopasture, as it can be applied to various climates, landscapes, and farming systems. These real-world examples inspire and guide those considering the implementation of silvopasture on their lands.

Ultimately, the widespread adoption of silvopasture and other agroforestry practices can significantly address global challenges such as climate change, biodiversity loss, and food security. Furthermore, by embracing silvopasture, we can work towards a more sustainable, resilient, and productive agricultural future for generations to come.

Chapter Summary

- Silvopasture is a sustainable agricultural practice that integrates livestock, trees, and forage into a single system, offering environmental, economic, and social benefits.
- Environmental benefits of silvopasture include carbon sequestration, increased biodiversity, and improved water conservation and quality.
- Economic benefits of silvopasture include diversified farm income, increased productivity, and reduced input costs for farmers.

- Social benefits of silvopasture include promoting food security and preserving cultural heritage in rural communities.
- Key components of a successful silvopasture system include careful selection of tree species, strategic tree planting and spacing, diverse and nutritious forage, adaptive livestock management, and ongoing monitoring and evaluation.
- Best practices for designing and implementing silvopasture include assessing the site, choosing the right tree species, integrating livestock, managing forage, protecting and maintaining trees, monitoring and evaluating the system, and investing in education and training.
- Real-world case studies demonstrate the versatility and effectiveness of silvopasture systems in various regions and climates, showcasing their potential to create more sustainable, productive, and resilient agricultural systems.
- The future of silvopasture is promising, as its widespread adoption can play a significant role in addressing global challenges such as climate change, biodiversity loss, and food security, contributing to a more sustainable and resilient agricultural future.

4

AGROFORESTRY SYSTEMS AND PRACTICES

A thriving agroforestry system, where diverse crops and trees are strategically integrated, creating a harmonious and sustainable agricultural landscape.

Agroforestry, a term coined in the *1970s*, is an age-old practice revitalized in recent years as a sustainable approach to agriculture and land management. At its core, agroforestry is the integration of trees, crops, and livestock into a single, ecologically balanced system. This harmonious coexistence of various elements not only enhances the productivity of the land but also provides numerous environmental, social, and economic benefits.

This chapter will explore the fascinating world of agroforestry systems and practices, exploring traditional and modern approaches. We will examine agroforestry's key components and benefits, with a particular focus on silvopasture – a unique practice that combines forestry and livestock management. Through case studies, we will showcase successful implementations of agroforestry and silvopasture around the world, and finally, we will discuss the prospects of these sustainable land management techniques.

As we embark on this journey, we must understand that agroforestry is not a one-size-fits-all solution. Instead, it is a highly adaptable practice that can be tailored to suit the specific needs and conditions of a given region, climate, and community. The beauty of agroforestry lies in its flexibility and its ability to foster a symbiotic relationship between humans and nature, ensuring the long-term health and prosperity of both.

So, let's begin our exploration of agroforestry systems and practices and discover how this ancient wisdom can help shape a more sustainable and resilient future for our planet and its inhabitants.

Traditional and Modern Agroforestry Systems

Agroforestry, the practice of integrating trees and shrubs with crops and livestock, has been a part of human agriculture for centuries. Traditional and modern agroforestry systems have evolved to meet the needs of diverse communities and environments, offering a range of benefits for both people and the planet. This section will explore the key characteristics and examples of traditional and modern agroforestry systems, highlighting their unique features and contributions to sustainable agriculture.

Traditional agroforestry systems have been shaped by generations of farmers who have skillfully combined their knowledge of local ecosystems with their agricultural needs. These systems are often characterized by a high degree of biodiversity, as farmers have learned to cultivate various plant and animal species that complement and support one another. Some examples of traditional agroforestry systems include:

Homegardens: These are found in many tropical and subtropical regions. Homegardens are small-scale, multi-layered agroforestry systems that incorporate diverse trees, shrubs, and crops. They often serve as a primary source of food, medicine, and income for rural households while providing valuable ecosystem services such as soil conservation and carbon sequestration.

Shifting cultivation: Also known as swidden agriculture or slash-and-burn, shifting cultivation is a traditional agroforestry practice in which farmers clear a small forest area, burn the vegetation, and then plant crops in the nutrient-rich ash. After a few years of cultivation, the land is left fallow to regenerate, and the process is repeated in a new area. This system allows for the sustainable use of forest resources while maintaining soil fertility and biodiversity.

Forest farming: In this system, farmers cultivate shade-tolerant crops such as coffee, cocoa, or spices under the canopy of a managed forest. This practice provides a sustainable source of income for farmers and helps conserve forest ecosystems and their associated biodiversity.

On the other hand, modern agroforestry systems have emerged in response to the challenges posed by industrial agriculture and the need for more sustainable land management practices. These systems often incorporate scientific knowledge and technological innovations to optimize productivity, resource use efficiency, and environmental benefits. Some examples of modern agroforestry systems include:

Alley cropping: This system involves planting rows of trees or shrubs alongside annual or perennial crops, creating a series of "alleyways" for cultivation. The trees provide shade, wind protection, and nutrient cycling, while the crops benefit from improved microclimates and reduced soil erosion. Common tree species used in alley cropping include nitrogen-fixing species such as Leucaena or Gliricidia, which can also provide fodder for livestock.

Riparian buffer strips: In this system, trees and shrubs are planted along watercourses to protect them from agricultural runoff and erosion. The vegetation acts as a natural filter, absorbing nutrients and sediment while providing wildlife habitat and stabilizing stream banks.

Windbreaks and shelterbelts: These linear plantings of trees and shrubs are designed to protect crops, livestock, and infrastructure from wind damage, reduce soil erosion, and conserve energy. Windbreaks can also enhance biodiversity by providing habitat for birds and other wildlife.

In conclusion, traditional and modern agroforestry systems offer diverse practices that can be adapted to suit different environmental and socio-economic contexts. Furthermore, by integrating trees and shrubs with crops and livestock, these systems contribute to sustainable agriculture, food security, and environmental conservation, making them essential to our efforts to build a more resilient and sustainable future.

Key Components and Benefits of Agroforestry

As an innovative and sustainable land management practice, agroforestry combines the cultivation of trees, crops, and livestock in a mutually beneficial system. This approach optimizes land use and provides numerous environmental, economic, and social benefits. In this section, we will delve into the critical components of agroforestry and explore the advantages it offers over conventional agricultural practices.

The foundation of agroforestry lies in its three primary components: **trees, crops, and livestock.** These elements interact with one another in various ways, creating a dynamic and resilient ecosystem. Trees provide shade, shelter, and habitat for wildlife, while their roots help stabilize the soil and prevent erosion. Crops, in turn, benefit from the improved soil structure and increased organic matter provided by the trees. Livestock can graze on the understory vegetation and contribute to nutrient cycling through their manure. The integration of these components results in a diverse and productive system that can withstand fluctuations in climate and market conditions.

The benefits of agroforestry are numerous and can be categorized into environmental, economic, and social advantages.

Environmental Benefits

1. **Soil Conservation:** Tree roots help stabilize the soil, reducing erosion and improving soil structure. Adding organic matter from leaf litter and root decomposition also enhances soil fertility.
2. **Biodiversity:** Agroforestry systems support a wide range of plant and animal species, promoting ecological balance and preserving habitats for beneficial insects, birds, and mammals.
3. **Carbon Sequestration:** Trees absorb and store carbon dioxide, helping to mitigate climate change. The integration of trees in agricultural landscapes can significantly increase the overall carbon sequestration potential of the land.
4. **Water Management:** Agroforestry practices can improve water infiltration and retention, reducing the risk of flooding and drought. Trees also act as natural filters, improving water quality by absorbing pollutants and nutrients.

Economic Benefits:

1. **Diversified Income:** Agroforestry systems offer multiple sources of income, including timber, non-timber forest products, and livestock. This diversification reduces the risk associated with market fluctuations and crop failures.
2. **Increased Productivity:** The complementary nature of agroforestry components can lead to higher overall yields. For example, shade-tolerant crops can benefit from the shelter trees provide, while nitrogen-fixing trees can improve soil fertility for neighboring crops.

3. **Cost Savings:** Agroforestry practices can reduce the need for chemical fertilizers, pesticides, and irrigation, lowering production costs and increasing profit margins.

Social Benefits

1. **Food Security:** The diversification of crops and livestock in agroforestry systems can improve food security and nutrition for local communities.
2. **Cultural Preservation:** Many traditional agroforestry practices are rooted in indigenous knowledge and culture, and their continued use helps preserve these valuable traditions.
3. **Community Development:** Agroforestry can provide opportunities for education, training, and employment, fostering social cohesion and empowering rural communities.

In conclusion, the key components of agroforestry – trees, crops, and livestock – work together to create a sustainable and resilient system that offers a multitude of environmental, economic, and social benefits. By embracing agroforestry practices, farmers and land managers can contribute to a more sustainable and productive future for agriculture and forestry.

Silvopasture: Integrating Livestock and Forestry

Silvopasture, a harmonious marriage of livestock and forestry, is a practice that has been gaining momentum in recent years. As a subcategory of agroforestry, silvopasture combines the cultivation of trees with the rearing of livestock, creating a mutually beneficial ecosystem. This innovative approach to land management not only optimizes the use of available resources but also offers a plethora of environmental, economic, and social benefits.

At the heart of silvopasture lies the concept of **symbiosis.** Trees provide shade, shelter, and forage for livestock, while the animals contribute to the health of the soil through their natural grazing and waste production. This dynamic relationship results in improved animal welfare, increased biodiversity, and enhanced soil fertility, making silvopasture a sustainable and efficient land management practice.

One of the key advantages of silvopasture is its ability to **diversify income streams** for farmers. Farmers can tap into multiple markets by integrating trees and livestock, such as timber, non-timber forest products, and animal products like meat, milk, and wool. This diversification not only increases the overall profitability of the farm but also provides a safety net in case one market faces challenges.

Another compelling aspect of silvopasture is its potential to mitigate climate change. Trees act as carbon sinks, absorbing and storing atmospheric carbon dioxide while providing shade that reduces the need for energy-intensive cooling systems for livestock. Moreover, the integration of trees and livestock can reduce greenhouse gas emissions, as improved soil health and fertility promote better nutrient cycling and reduce the need for synthetic fertilizers.

Silvopasture also plays a vital role in promoting biodiversity. By creating a varied landscape with different plant species, habitats, and microclimates, silvopasture systems encourage the growth of diverse flora and fauna. This increased biodiversity benefits the ecosystem's overall health and provides essential ecosystem services such as pollination, pest control, and nutrient cycling.

Implementing silvopasture, however, is not without its challenges. It requires careful planning, management, and monitoring to maintain the balance between trees and livestock. Farmers must consider factors such as tree species selection, planting density, and grazing patterns to create a productive and sustainable system.

In conclusion, silvopasture is a promising and innovative land management approach with many benefits. By integrating livestock and forestry, this practice optimizes the use of available resources and contributes to environmental conservation, economic stability, and social well-being. As we look toward the future of agroforestry and silvopasture, it is essential to continue exploring and refining these practices to ensure their long-term success and sustainability.

Case Studies: Successful Agroforestry and Silvopasture Implementations

In this section, we will delve into real-life examples of successful agroforestry and silvopasture implementations from around the world. These case studies showcase the potential of these systems to enhance agricultural productivity, promote biodiversity, and contribute to sustainable land management practices.

Case Study 1: The Chagga Home Gardens of Tanzania

The Chagga people of Mount Kilimanjaro in Tanzania have practiced agroforestry for centuries. Their home gardens, known as *"vihamba,"* are an excellent example of traditional agroforestry systems. These gardens consist of a diverse mix of trees, shrubs, and crops, including coffee, bananas, and various fruit trees. The multi-layered canopy structure provides shade for the understory crops, while the trees help to conserve soil moisture and prevent erosion. The Chagga home gardens are not only productive but also support a high level of biodiversity, contributing to the overall resilience of the system.

Case Study 2: The Quesungual System in Honduras

In the dry and degraded hillsides of western Honduras, farmers have developed an innovative agroforestry system known as the Quesungual system. This system involves the selective pruning of native trees to create a sparse canopy, allowing sunlight to reach the understory crops such as maize, beans, and sorghum. The pruned branches are used as mulch, which helps to conserve soil moisture and improve soil fertility. The Quesungual system has been shown to increase crop yields, reduce soil erosion, and enhance biodiversity compared to conventional agricultural practices in the region.

Case Study 3: Silvopasture in the Dehesa of Spain

The Dehesa is a traditional silvopastoral system found in the Iberian Peninsula, characterized by the integration of livestock grazing with oak tree forests. The oak trees provide shade and acorns for the grazing animals, such as the Iberian pig and Merino sheep, while the livestock help to maintain the grasslands and promote tree regeneration. This symbiotic relationship results in a sustainable and productive system supporting diverse flora and fauna. The Dehesa is also known for its high-quality products, such as the famous *Iberian ham*, which are highly valued in local and international markets.

Case Study 4: Alley Cropping in the United States

Alley cropping is an agroforestry practice that involves planting rows of trees or shrubs alongside annual or perennial crops. This system has been successfully implemented in various regions across the United States, providing multiple benefits such as improved soil fertility, erosion control, and increased crop yields. One notable example is a farm in Missouri, where a farmer has integrated walnut trees with soybean and maize crops. The walnut trees provide valuable timber and nuts, while the soybean and maize crops benefit from the nitrogen-fixing properties of the trees and the improved microclimate created by the tree canopy.

In conclusion, these case studies demonstrate the versatility and adaptability of agroforestry and silvopasture systems in different ecological and socio-economic contexts. By learning from these successful implementations, we can continue to promote and develop sustainable land management practices that benefit both people and the environment.

The Future of Agroforestry and Silvopasture

As we reach the end of this enlightening journey through the world of agroforestry and silvopasture, it is essential to reflect on the immense potential these practices hold for the future of agriculture, forestry, and environmental conservation. Integrating trees, crops, and livestock in a single system has proven to be a sustainable and productive approach to land management, offering myriad benefits for farmers, communities, and the environment.

The future of agroforestry and silvopasture is undoubtedly promising. As the global population grows, the demand for food, fiber, and fuel will only increase. Agroforestry systems have the potential to meet these demands while simultaneously addressing the challenges of climate change, biodiversity loss, and soil degradation. By adopting these practices, farmers can diversify their income sources, reduce their vulnerability to market fluctuations, and improve their overall resilience in the face of environmental and economic uncertainties.

Furthermore, the benefits of agroforestry and silvopasture extend beyond the farm gate. These systems can play a crucial role in mitigating climate change by sequestering carbon, reducing greenhouse gas emissions, and promoting renewable energy sources. Additionally, they can contribute to biodiversity conservation by providing a habitat for various species of flora and fauna and maintaining the ecological balance of the landscape.

In order to harness the full potential of agroforestry and silvopasture, it is essential to invest in research, education, and extension services. This will enable farmers to access the knowledge, skills, and resources necessary to implement these systems effectively and efficiently. Moreover, supportive policies and incentives should be implemented to encourage the widespread adoption of agroforestry and silvopasture practices.

In conclusion, agroforestry and silvopasture represent a paradigm shift in the way we approach agriculture and forestry. By embracing these systems, we can cultivate a more sustainable, resilient, and productive future for our planet and its inhabitants. The journey may be long and challenging, but the rewards are well worth the effort. So, as we move forward, let us continue to **explore, innovate, and collaborate** in our quest to create a greener, more prosperous world through the power of agroforestry and silvopasture.

Chapter Summary

- Agroforestry is the integration of trees, crops, and livestock into a single, ecologically balanced system, offering numerous environmental, social, and economic benefits.
- Traditional and modern agroforestry systems have evolved to meet the needs of diverse communities and environments. Examples include homegardens, shifting cultivation, forest farming, alley cropping, riparian buffer strips, and windbreaks.
- The key components of agroforestry are trees, crops, and livestock, which interact with one another to create a dynamic and resilient ecosystem.
- Environmental benefits of agroforestry include soil conservation, biodiversity, carbon sequestration, and water management.
- Economic benefits of agroforestry include diversified income, increased productivity, and cost savings.
- Social benefits of agroforestry include food security, cultural preservation, and community development.
- Silvopasture is a subcategory of agroforestry that combines the cultivation of trees with the rearing of livestock, offering benefits such as diversification of income streams, climate change mitigation, and promotion of biodiversity.
- The future of agroforestry and silvopasture is promising, with the potential to address global challenges such as food security, climate change, and biodiversity loss, but it requires investment in research, education, and supportive policies to encourage widespread adoption.

5

THE ROLE OF AGROFORESTRY IN CLIMATE CHANGE MITIGATION

An image that showcases the harmonious coexistence of agriculture and forestry in a futuristic landscape, where agroforestry practices play a vital role in combating climate change and preserving biodiversity.

Climate change has emerged as one of the most pressing challenges of our time, with far-reaching consequences for ecosystems, economies, and societies worldwide. As global temperatures continue to rise, the need for innovative and sustainable solutions to mitigate the impacts of climate change becomes increasingly urgent. One such solution lies in the realm of agroforestry, a land management practice that integrates trees and shrubs with crops and livestock systems. This chapter explores the multifaceted role of agroforestry in climate change mitigation, delving into its potential to sequester carbon, enhance biodiversity, promote sustainable land management, and provide socioeconomic benefits to communities.

The interconnectedness of agroforestry and climate change is evident in the numerous ways that these systems can influence one another. On the one hand, climate change can exacerbate existing challenges faced by agricultural systems, such as soil degradation, water scarcity, and reduced crop yields. On the other hand, agroforestry practices can help to address these challenges by enhancing the resilience of ecosystems and providing a range of ecosystem services that contribute to climate change mitigation and adaptation. By integrating trees and shrubs into agricultural landscapes, agroforestry systems can create more diverse and resilient ecosystems that are better equipped to withstand the impacts of a changing climate.

At the heart of agroforestry's potential to mitigate climate change is its **capacity for carbon sequestration**, or the process by which trees and other plants capture and store atmospheric carbon dioxide. As trees grow, they absorb carbon dioxide from the atmosphere and convert it into biomass, effectively removing this potent greenhouse gas from the atmosphere and reducing its contribution to global warming. In addition to carbon sequestration, agroforestry systems can also enhance biodiversity and ecosystem resilience, providing habitats for a wide range of plant and animal species and promoting the overall health and stability of ecosystems.

Sustainable land management practices are another key component of agroforestry's role in climate change mitigation. By adopting practices such as conservation tillage, cover cropping, and agroforestry, farmers can improve soil health, reduce erosion, and increase water infiltration, all of which contribute to the overall resilience of agricultural systems in the face of climate change. Furthermore, these practices can also help to reduce the need for synthetic fertilizers and pesticides, which are significant sources of greenhouse gas emissions.

The socioeconomic benefits of agroforestry in a changing climate should not be overlooked. By diversifying agricultural systems and providing additional sources of income through the sale of timber, non-timber forest products, and ecosystem services, agroforestry can help to improve the livelihoods of rural communities and enhance their capacity to adapt to the impacts of climate change. Moreover, agroforestry can also contribute to food security by providing a diverse array of nutritious food sources, such as fruits, nuts, and edible leaves.

In conclusion, the interconnectedness of agroforestry and climate change underscores the importance of embracing this land management practice as a pathway to a sustainable future. By harnessing the power of agroforestry to sequester carbon, enhance biodiversity, promote sustainable land management, and provide socioeconomic benefits, we can work towards mitigating the impacts of climate change and building more resilient agricultural systems for generations to come.

Carbon Sequestration: A Natural Solution to Greenhouse Gas Emissions

As we delve into the heart of the matter, it is crucial to understand the concept of carbon sequestration and its significance in mitigating climate change. Simply put, carbon sequestration is the process by which carbon dioxide (CO_2) is captured and stored, preventing it from being released into the atmosphere and contributing to global warming.

The beauty of agroforestry lies in its ability to **sequester carbon naturally**, harnessing the power of trees and vegetation to combat greenhouse gas emissions.

Trees are the unsung heroes in the fight against climate change. Through the process of photosynthesis, they absorb CO_2 from the atmosphere and convert it into oxygen and biomass. This biomass, composed of roots, stems, branches, and leaves, stores carbon, effectively removing it from the atmosphere. Agroforestry systems, which integrate trees with crops and livestock, create a synergistic environment that maximizes carbon sequestration potential.

The benefits of carbon sequestration in agroforestry systems extend beyond the mere absorption and storage of CO2. By incorporating trees into agricultural landscapes, farmers can create microclimates that help regulate temperature and humidity, reducing the need for energy-intensive irrigation and cooling systems. Moreover, the shade trees provide can reduce soil evaporation, conserve water and maintain soil moisture levels.

In addition to the direct impact on carbon sequestration, agroforestry practices indirectly contribute to climate change mitigation. For instance, using nitrogen-fixing trees can improve soil fertility, reducing the need for synthetic fertilizers that emit nitrous oxide, a potent greenhouse gas. Furthermore, agroforestry systems can prevent soil erosion and degradation, preserving the soil's capacity to store carbon.

The potential of agroforestry to sequester carbon is immense. According to the Intergovernmental Panel on Climate Change (IPCC), agroforestry practices can sequester between 0.4 and 1.2 gigatons of CO2 per year, making it one of the most promising natural climate solutions. However, it is essential to recognize that the effectiveness of carbon sequestration in agroforestry systems depends on various factors, such as tree species, climate, and management practices.

In conclusion, carbon sequestration through agroforestry offers a natural and sustainable solution to mitigate greenhouse gas emissions. By integrating trees into agricultural systems, we can harness their power to absorb and store carbon, regulate microclimates, and improve soil health. As we continue to explore the role of agroforestry in climate change mitigation, it becomes increasingly evident that embracing these practices is beneficial for the environment and the well-being of future generations.

Enhancing Biodiversity and Ecosystem Resilience in Agroforestry Systems

The beauty of agroforestry lies in its ability to create a harmonious balance between agricultural production and environmental conservation. One of the most significant advantages of agroforestry systems is their potential to enhance biodiversity and bolster ecosystem resilience. This section will delve into the ways in which agroforestry contributes to the preservation and promotion of biodiversity, as well as the strengthening of ecosystems in the face of climate change.

Biodiversity, or the variety of life in a particular ecosystem, is crucial to a healthy environment. It ensures the **stability and productivity of ecosystems**, which in turn provide essential services such as food, water, and climate regulation. By their very nature, agroforestry systems promote biodiversity by integrating a diverse range of plant and animal species into agricultural landscapes. This integration creates a mosaic of habitats that support a wide array of flora and fauna, including pollinators, predators, and decomposers, all of which play vital roles in maintaining ecosystem health.

In addition to supporting a diverse range of species, agroforestry systems foster genetic diversity within individual species. This is particularly important for crop plants, as genetic diversity provides the raw material for adaptation and evolution in response to changing environmental conditions. By incorporating a variety of crop species and cultivars into agroforestry systems, farmers can create a more resilient agricultural landscape that is better equipped to cope with the challenges posed by climate change.

Ecosystem resilience, or the capacity of an ecosystem to recover from disturbances and maintain its essential functions, is another critical aspect of environmental health. Agroforestry systems can **enhance ecosystem resilience** in several ways. First, by promoting biodiversity, agroforestry systems create a more robust and adaptable ecosystem that can better withstand disturbances such as pests, diseases, and extreme weather events. The presence of multiple species and functional groups within an agroforestry system can also help buffer against the loss of individual species, ensuring that the ecosystem remains functional and productive.

Second, agroforestry systems can help **maintain and improve soil health**, a key factor in ecosystem resilience. Integrating trees and shrubs into agricultural landscapes can reduce soil erosion, improve soil structure, and increase the availability of essential nutrients. These benefits, in turn, promote the growth of a diverse and healthy soil microbial community, which plays a crucial role in nutrient cycling, decomposition, and overall ecosystem functioning.

Lastly, agroforestry systems can contribute to **landscape-scale resilience** by creating connectivity between fragmented habitats. This connectivity allows for the movement of species between patches of suitable habitat, which is essential for maintaining genetic diversity and ensuring the long-term survival of populations. In this way, agroforestry systems can serve as important stepping stones for wildlife in increasingly fragmented landscapes, helping to maintain the ecological integrity of the broader environment.

In conclusion, integrating agroforestry systems into agricultural landscapes offers a promising pathway for enhancing biodiversity and ecosystem resilience in the face of climate change. By fostering a diverse and interconnected network of species and habitats, agroforestry systems can help create a more robust and adaptable environment better equipped to withstand the challenges of a changing climate. As we continue to grapple with the impacts of climate change on our planet, we must embrace and promote agroforestry as a vital component of a sustainable and resilient future.

Sustainable Land Management Practices for Climate Change Adaptation

As the world grapples with the ever-increasing impacts of climate change, it has become imperative to adopt sustainable land management practices that promote climate change adaptation. With its unique blend of agriculture and forestry, agroforestry offers many solutions to help farmers and communities adapt to the changing climate while maintaining productivity and ecosystem health. In this section, we will delve into some of the sustainable land management practices that can be employed in agroforestry systems to **enhance climate change adaptation.**

First and foremost, the selection of appropriate tree species and crop varieties is crucial in building a resilient agroforestry system. By choosing species that are well-adapted to local conditions and resistant to pests and diseases, farmers can ensure the long-term productivity and stability of their agroforestry systems. Moreover, incorporating a diverse mix of species can further enhance resilience, as it reduces the risk of catastrophic losses due to climate-related events such as droughts, floods, or pest outbreaks.

Another key practice in sustainable land management is the conservation of soil and water resources. Agroforestry systems can help in this regard by reducing soil erosion, improving soil structure, and increasing water infiltration and retention. For instance, the strategic planting of trees and shrubs along contour lines or on slopes can act as a barrier against soil erosion, while the deep root systems of trees can help stabilize the soil and improve its structure. Additionally, the presence of trees can enhance water infiltration and retention, reducing the impacts of droughts and floods on crops and local water resources.

Incorporating agroforestry practices that promote natural pest control is another way to enhance climate change adaptation. By fostering a diverse and balanced ecosystem, agroforestry systems can encourage the presence of beneficial insects and other natural predators, reducing the need for chemical pesticides and enhancing the system's overall resilience. For example, integrating insectary plants or creating a habitat for natural predators such as birds and bats can help keep pest populations in check.

Furthermore, agroforestry systems can be designed to optimize the use of available resources, such as sunlight, water, and nutrients, through the strategic arrangement of trees, crops, and livestock. This can be achieved by employing practices such as **alley cropping**, where rows of trees are interspersed with rows of crops, or by integrating livestock into the system to recycle nutrients and reduce the need for external inputs.

By making the most of these resources, agroforestry systems can maintain productivity even under changing climate conditions.

Lastly, engaging local communities and farmers in the planning, implementation, and management of agroforestry systems is essential for their long-term success and sustainability. By involving stakeholders in the decision-making process and providing them with the necessary knowledge and tools, agroforestry can empower communities to take charge of their own climate change adaptation strategies.

In conclusion, sustainable land management practices play a vital role in enhancing the adaptive capacity of agroforestry systems in the face of climate change. By incorporating practices that promote resilience, resource conservation, and community engagement, agroforestry can be a powerful tool in our collective efforts to adapt to a changing climate and secure a sustainable future for generations to come.

The Socioeconomic Benefits of Agroforestry in a Changing Climate

As we delve deeper into the multifaceted world of agroforestry, it becomes increasingly evident that this sustainable land management practice offers far more than just environmental benefits. In a world grappling with the consequences of climate change, agroforestry also presents a myriad of socioeconomic advantages that can help communities adapt and thrive in the face of these challenges. In this section, we will explore the various ways in which agroforestry contributes to the socioeconomic well-being of communities and nations alike.

First and foremost, agroforestry systems provide a diversified source of income for farmers and landowners. By integrating trees and shrubs with traditional crops and livestock, agroforestry allows for the simultaneous production of various marketable goods, such as timber, fruits, nuts, and medicinal plants. This diversification not only reduces the risks associated with monoculture farming but also creates new income-generating opportunities for rural communities. Moreover, the integration of high-value tree crops can significantly increase the overall profitability of farming systems, leading to improved livelihoods and economic stability.

In addition to boosting income, agroforestry systems contribute to **food security** by enhancing the availability, access, and utilization of diverse and nutritious food sources. Trees and shrubs integrated into agricultural landscapes can provide a variety of edible products, such as fruits, nuts, and leaves, which can supplement and diversify the diets of local communities. Furthermore, agroforestry practices can improve soil fertility and water availability, thereby increasing the productivity of adjacent food crops. In the face of climate change-induced disruptions to food systems, agroforestry's role in bolstering food security becomes all the more crucial.

Agroforestry also plays a significant role in fostering social cohesion and community resilience. By promoting the sustainable use of natural resources, agroforestry can help reduce conflicts over land and water, which are often exacerbated by climate change. Additionally, the implementation of agroforestry projects often involves collaboration among various stakeholders, including farmers, government agencies, non-governmental organizations, and research institutions. This collaborative approach fosters a sense of community ownership and shared responsibility, **strengthening social bonds** and enhancing the collective capacity to adapt to climate change.

Another significant socioeconomic benefit of agroforestry is its potential to create **employment opportunities**, particularly in rural areas. The labor-intensive nature of agroforestry practices, such as planting, pruning, and harvesting, generates jobs for local communities, thereby contributing to poverty alleviation and rural development. Moreover, the processing and marketing of agroforestry products can stimulate the growth of local industries and value chains, further enhancing economic opportunities and regional development.

In conclusion, the socioeconomic benefits of agroforestry in a changing climate are manifold and far-reaching. By diversifying income sources, enhancing food security, fostering social cohesion, and creating employment opportunities, agroforestry systems can significantly contribute to the well-being and resilience of communities in the face of climate change. As we strive to build a more sustainable and equitable future, embracing agroforestry as a holistic land management practice becomes a viable option and a necessary step toward achieving our collective goals.

Embracing Agroforestry as a Pathway to a Sustainable Future

In conclusion, it is evident that agroforestry and silvopasture play a significant role in mitigating the impacts of climate change while providing numerous environmental, social, and economic benefits. Therefore, as our world grapples with the consequences of a rapidly changing climate, we must explore and embrace sustainable land management practices such as agroforestry to ensure a resilient and thriving future for our planet and its inhabitants.

Agroforestry systems have demonstrated their potential to sequester carbon, enhance biodiversity, and promote ecosystem resilience. By integrating trees and shrubs with crops and livestock, these systems not only contribute to climate change mitigation but also provide a diverse array of ecosystem services that support human well-being and the **health of our planet.** Furthermore, agroforestry practices can help farmers adapt to the changing climate by improving soil fertility, reducing erosion, and providing shade and shelter for crops and animals.

The socioeconomic benefits of agroforestry are also noteworthy, as these systems can improve livelihoods, create job opportunities, and contribute to food security. Furthermore, by diversifying agricultural production and income sources, agroforestry can help farmers and rural communities become more resilient to the uncertainties of climate change and market fluctuations. Additionally, agroforestry can promote social equity and gender equality by empowering women and marginalized groups to participate in decision-making processes and benefit from the resources and opportunities provided by these systems.

As we move forward, policymakers, researchers, and practitioners must work together to promote the widespread adoption of agroforestry and silvopasture systems. This can be achieved through the development of supportive policies, increased investment in research and development, and the provision of technical assistance and training to farmers. Moreover, fostering collaboration and knowledge exchange among stakeholders, including farmers, researchers, and policymakers, is crucial to ensure that agroforestry practices are tailored to local contexts and address the specific needs and challenges of different communities.

In the face of climate change, embracing agroforestry as a pathway to a sustainable future is not only a wise decision but a necessary one. By integrating trees into our agricultural landscapes, we can harness the power of nature to **mitigate climate change**, enhance ecosystem resilience, and improve the well-being of people and the planet. As we continue to confront the challenges posed by a changing climate, let us remember that the solutions we seek may already be rooted in the ground beneath our feet.

Chapter Summary

- Agroforestry, a land management practice that integrates trees and shrubs with crops and livestock systems, plays a crucial role in climate change mitigation and adaptation.
- Agroforestry systems have the potential to sequester significant amounts of carbon, reducing greenhouse gas emissions and contributing to climate change mitigation.
- By enhancing biodiversity and ecosystem resilience, agroforestry systems create more diverse and adaptable ecosystems that can better withstand the impacts of climate change.
- Sustainable land management practices, such as conservation tillage, cover cropping, and agroforestry, can improve soil health, reduce erosion, and increase water infiltration, contributing to climate change adaptation.
- Agroforestry systems provide socioeconomic benefits, such as diversified income sources, improved livelihoods, and enhanced food security, helping communities adapt to climate change.
- The implementation of agroforestry practices can foster social cohesion and community resilience by promoting the

sustainable use of natural resources and encouraging collaboration among stakeholders.

- Policymakers, researchers, and practitioners must work together to promote the widespread adoption of agroforestry systems through supportive policies, increased investment in research and development, and technical assistance and training for farmers.
- Embracing agroforestry as a pathway to a sustainable future is essential in addressing the challenges posed by climate change, as it harnesses the power of nature to mitigate climate change, enhance ecosystem resilience, and improve the well-being of people and the planet.

6

BIODIVERSITY CONSERVATION AND ECOSYSTEM SERVICES IN AGROFORESTRY

An image that showcases a thriving agroforestry landscape, complete with diverse plant and animal species coexisting harmoniously.

In today's rapidly changing world, the need for sustainable agricultural practices that promote biodiversity conservation and provide essential ecosystem services has never been more critical. Agroforestry, the intentional integration of trees and shrubs into crop and animal farming systems, offers a promising solution to this challenge. By combining the best of both worlds - agriculture and forestry - agroforestry systems can enhance biodiversity, improve soil health, and contribute to climate change mitigation, among other benefits.

This chapter delves into the vital role of agroforestry and silvopasture in **biodiversity conservation** and the **provision of ecosystem services.** We will begin by exploring the importance of biodiversity in agroforestry systems and how these systems can contribute to the conservation of various plant and animal species. Next, we will discuss the numerous ecosystem services that agroforestry provides, including carbon sequestration, soil fertility enhancement, and water regulation.

Silvopasture, a specific type of agroforestry that combines livestock grazing with the cultivation of trees, will be examined as a sustainable approach to biodiversity conservation. Finally, we will conclude by discussing the future of biodiversity conservation and ecosystem services in agroforestry, highlighting the potential of these systems to address some of the most pressing environmental challenges of our time.

As you read through this chapter, we invite you to consider the immense potential of agroforestry and silvopasture as tools for biodiversity conservation and ecosystem services provision. By understanding the benefits and challenges of these practices, we can work together to create a more sustainable, resilient, and biodiverse world for future generations.

The Role of Agroforestry in Enhancing Biodiversity

Agroforestry, the intentional integration of trees and shrubs into crop and animal farming systems, plays a vital role in enhancing biodiversity. By creating a more diverse and complex landscape, agroforestry systems promote a wide range of flora and fauna, contributing to the overall health and resilience of ecosystems. In this section, we will explore the various ways in which agroforestry contributes to biodiversity conservation and the benefits that arise from this symbiotic relationship.

One of the most significant ways in which agroforestry enhances biodiversity is by **providing habitat for various species.** Trees and shrubs offer shelter, nesting sites, and food sources for numerous birds, mammals, insects, and other organisms. This, in turn, creates a more balanced ecosystem, as the presence of diverse species helps to regulate pests and diseases, reducing the need for chemical interventions. Moreover, the integration of trees and shrubs within agricultural landscapes can serve as biological corridors, connecting fragmented habitats and allowing for the movement of species across the landscape.

Agroforestry systems also contribute to the conservation of genetic diversity, which is crucial for the long-term adaptability and survival of species. By cultivating various crops and tree species, agroforestry practices help preserve the genetic resources of both cultivated and wild plants. This is particularly important in the face of climate change, as genetic diversity enables species to adapt to changing environmental conditions and withstand various stressors.

In addition to conserving biodiversity, agroforestry systems also promote the **abundance and diversity of pollinators**, such as bees, butterflies, and other insects. Pollinators are essential for the reproduction of many plant species, including those grown for food, and their decline has been a growing concern worldwide. By providing diverse floral resources and nesting sites, agroforestry systems support healthy pollinator populations, which in turn contribute to increased crop yields and overall ecosystem health.

Furthermore, agroforestry practices can help to maintain and enhance soil biodiversity. The presence of trees and shrubs in agricultural systems promotes the growth of diverse microorganisms, such as bacteria and fungi, which play a crucial role in nutrient cycling and soil fertility. These microorganisms, along with other soil-dwelling organisms like earthworms, contribute to the breakdown of organic matter and the release of essential nutrients, ultimately supporting the growth of healthy crops.

In conclusion, agroforestry plays a critical role in enhancing biodiversity by providing habitat for various species, conserving genetic diversity, supporting pollinator populations, and maintaining soil biodiversity. The benefits of this increased biodiversity are numerous; they include improved ecosystem health, pest and disease regulation, and increased crop yields. Farmers and land managers can contribute to biodiversity conservation by adopting agroforestry practices while reaping the rewards of a more resilient and productive agricultural system.

Ecosystem Services Provided by Agroforestry Systems

Agroforestry systems, which integrate trees, shrubs, and crops, provide a wide range of ecosystem services that benefit both humans and the environment. These services are the direct and indirect contributions of ecosystems to human well-being, and they play a crucial role in sustaining the health and productivity of the planet. In this section, we will explore the various ecosystem services offered by agroforestry systems, emphasizing their importance in maintaining a balanced and thriving ecosystem.

First and foremost, agroforestry systems contribute to climate regulation by **sequestering carbon dioxide** from the atmosphere. Trees and other vegetation in these systems absorb and store carbon in their biomass, helping to mitigate the effects of climate change. Moreover, agroforestry practices can reduce greenhouse gas emissions from agriculture by decreasing the need for synthetic fertilizers and other chemical inputs.

Another vital ecosystem service that agroforestry systems provide is the **enhancement of soil fertility and quality.** Trees and shrubs in agroforestry systems improve soil structure by reducing erosion, increasing organic matter content, and promoting nutrient cycling. The deep root systems of trees also help to access nutrients from lower soil layers, making them available to nearby crops. Additionally, some tree species, such as legumes, can fix atmospheric nitrogen, further enriching the soil and reducing the need for synthetic fertilizers.

Agroforestry systems also contribute to **water regulation and purification.** The presence of trees and shrubs helps to reduce surface runoff, increase water infiltration, and recharge groundwater resources. This, in turn, prevents soil erosion, reduces the risk of flooding, and maintains water quality by filtering pollutants and sediment. Furthermore, agroforestry practices can enhance water use efficiency by providing shade and reducing evapotranspiration, which is particularly important in arid and semi-arid regions.

Biodiversity conservation is another significant ecosystem service offered by agroforestry systems. By creating a diverse and complex habitat, agroforestry practices support flora and fauna, pollinators, natural predators, and other beneficial organisms. This increased biodiversity can improve crop pollination, enhance pest control, and overall greater agroecosystem resilience.

Agroforestry systems also provide various provisioning services, such as food, fuel, and timber. By integrating trees and shrubs with crops, these systems can produce diverse products, including fruits, nuts, vegetables, and fodder for livestock. This diversification can improve food security and nutrition and provide additional income sources for farmers. Moreover, agroforestry practices can supply sustainable sources of fuelwood and timber, reducing pressure on natural forests and contributing to their conservation.

In conclusion, agroforestry systems offer many ecosystem services essential for the well-being of humans and the environment. By integrating trees, shrubs, and crops, these systems promote climate regulation, soil fertility, water management, biodiversity conservation, and the provision of various goods. As such, agroforestry practices represent a sustainable and resilient approach to land management, with the potential to address many of the challenges faced by modern agriculture.

Silvopasture: A Sustainable Approach to Biodiversity Conservation

Silvopasture, a unique and innovative agroforestry practice, combines the cultivation of trees, forage, and livestock in a single, integrated system. This sustainable approach optimizes land use and plays a vital role in biodiversity conservation. In this section, we will explore the various ways in which silvopasture contributes to the preservation of biodiversity and the overall health of ecosystems.

At its core, silvopasture is designed to **mimic natural ecosystems**, where multiple species coexist and interact harmoniously. By incorporating a diverse range of tree species, forage plants, and livestock breeds, silvopasture systems promote a rich and varied habitat that supports a wide array of flora and fauna. This diversity is essential for maintaining the ecological balance and resilience of the system, as it reduces the risk of disease outbreaks, pest infestations, and other threats to biodiversity.

One of the most significant benefits of silvopasture is its ability to **provide habitat for various wildlife species.** The integration of trees and shrubs within grazing lands creates a more complex and heterogeneous landscape, offering shelter, food, and nesting sites for birds, mammals, and insects. This, in turn, increases the overall species richness and abundance within the system, contributing to the preservation of local and regional biodiversity.

In addition to providing habitat for wildlife, silvopasture systems also contribute to biodiversity conservation by **enhancing the connectivity of fragmented landscapes.** As agricultural expansion and urbanization continue to encroach upon natural habitats, the creation of silvopasture systems can serve as vital corridors for wildlife movement between isolated patches of forest or grassland. This connectivity is crucial for the long-term survival of many species, as it allows for the exchange of genetic material and the recolonization of vacant habitats.

Moreover, silvopasture systems are crucial in **maintaining and enhancing soil biodiversity.** The presence of trees and shrubs within grazing lands helps to prevent soil erosion, improve soil structure, and increase the availability of nutrients. These factors, in turn, support a diverse and active community of soil organisms, including bacteria, fungi, and invertebrates. These organisms play a critical role in nutrient cycling, decomposition, and other essential ecosystem processes, ultimately contributing to the overall health and productivity of the system.

In conclusion, silvopasture represents a sustainable and promising approach to biodiversity conservation within agricultural landscapes. By integrating trees, forage, and livestock in a single system, silvopasture not only optimizes land use but also promotes a rich and varied habitat that supports a wide array of species. Furthermore, as the global demand for food continues to grow, adopting silvopasture practices can help ensure the long-term health and resilience of our ecosystems while also providing valuable ecosystem services to farmers and society as a whole.

The Future of Biodiversity Conservation and Ecosystem Services in Agroforestry

In conclusion, the future of biodiversity conservation and ecosystem services in agroforestry is promising and full of potential. As the world grapples with the challenges of climate change, food security, and environmental degradation, agroforestry and silvopasture practices offer sustainable solutions that can help address these pressing issues. By integrating trees, crops, and livestock, these systems can enhance biodiversity, improve soil health, and provide a range of ecosystem services that benefit both people and the environment.

The role of agroforestry in enhancing biodiversity cannot be overstated. By providing diverse habitats and resources for a wide array of species, agroforestry systems can support the conservation of flora and fauna. This, in turn, contributes to the overall health and resilience of ecosystems, which is crucial in the face of ongoing environmental challenges.

Ecosystem services provided by agroforestry systems are also of immense value. These services include carbon sequestration, water regulation, erosion control, and pollination, among others. By harnessing these benefits, agroforestry can help mitigate climate change, protect water resources, and support agricultural productivity. Furthermore, these services have the potential to generate economic value for farmers and landowners, providing incentives for the adoption of sustainable practices.

Silvopasture, as a specific form of agroforestry, offers a sustainable approach to biodiversity conservation by integrating livestock with trees and pasture. This practice can help maintain and restore degraded landscapes, reduce pressure on natural habitats, and support the conservation of species that depend on these ecosystems. Moreover, silvopasture can contribute to the livelihoods of farmers and rural communities by diversifying income sources and enhancing the resilience of agricultural systems.

The case studies presented in this chapter demonstrate the effectiveness of agroforestry and silvopasture practices in promoting biodiversity conservation and delivering ecosystem services. These examples inspire and guide the development and implementation of similar practices in different contexts around the world.

Looking ahead, the future of biodiversity conservation and ecosystem services in agroforestry depends on the continued research, innovation, and dissemination of knowledge in this field. Policymakers, researchers, and practitioners must work together to develop supportive policies, incentives, and educational programs that encourage the widespread adoption of agroforestry and silvopasture practices. By doing so, we can ensure that these systems play a central role in addressing the global challenges of our time and contribute to a more sustainable and resilient future for our planet.

Chapter Summary

- Agroforestry, the intentional integration of trees and shrubs into crop and animal farming systems, offers a sustainable solution to promote biodiversity conservation and provide essential ecosystem services.
- Agroforestry enhances biodiversity by providing habitat for various species, conserving genetic diversity, supporting pollinator populations, and maintaining soil biodiversity.
- Ecosystem services provided by agroforestry systems include climate regulation through carbon sequestration, soil fertility enhancement, water regulation and purification, biodiversity conservation, and the provision of various goods such as food, fuel, and timber.

- Silvopasture, a specific type of agroforestry that combines livestock grazing with the cultivation of trees, is a sustainable approach to biodiversity conservation that mimics natural ecosystems and promotes a rich and varied habitat.
- The future of biodiversity conservation and ecosystem services in agroforestry is promising, with the potential to address pressing global challenges such as climate change, food security, and environmental degradation.
- Continued research, innovation, and dissemination of knowledge in agroforestry and silvopasture are essential for their widespread adoption and success in promoting biodiversity conservation and delivering ecosystem services.
- Policymakers, researchers, and practitioners must collaborate to develop supportive policies, incentives, and educational programs that encourage the adoption of agroforestry and silvopasture practices, contributing to a more sustainable and resilient future for our planet.

7

SOCIOECONOMIC BENEFITS OF AGROFORESTRY AND SILVOPASTURE

A futuristic city skyline with agroforestry and silvopasture systems integrated into the urban environment.

In recent years, the world has witnessed a growing interest in sustainable agricultural practices that not only ensure food security but also contribute to the socioeconomic development of rural communities. Among these practices, agroforestry and silvopasture have emerged as promising solutions to address the challenges of climate change, land degradation, and rural poverty. By integrating trees, crops, and livestock on the same land, these systems offer a multitude of benefits that can significantly improve the livelihoods of millions of people, particularly in developing countries.

Agroforestry, which combines the cultivation of trees with crops, and silvopasture, which integrates trees with livestock grazing, have been practiced for centuries in various parts of the world. However, their potential to contribute to socioeconomic development has only recently been recognized and appreciated. This chapter aims to explore the various ways in which agroforestry and silvopasture can enhance the well-being of individuals, communities, and nations by fostering income diversification, poverty reduction, food security, employment opportunities, and environmental services.

The following sections will delve into the specific socioeconomic benefits of agroforestry and silvopasture practices, highlighting their potential to transform the lives of farmers and rural communities. We will also discuss the role of these systems in addressing global challenges such as climate change and land degradation and how they can contribute to the achievement of the United Nations Sustainable Development Goals (SDGs).

By the end of this chapter, readers will gain a deeper understanding of the growing importance of agroforestry and silvopasture in socioeconomic development and how these practices can be harnessed to create a more sustainable, equitable, and prosperous world for all.

Income Diversification and Poverty Reduction

In recent years, agroforestry and silvopasture practices have emerged as viable and sustainable alternatives to conventional agriculture, offering significant socioeconomic benefits to rural communities. One of the most notable advantages of these integrated land management systems is the potential for income diversification and poverty reduction. By combining the cultivation of trees, crops, and livestock, farmers can tap into **multiple revenue streams**, thereby reducing their vulnerability to market fluctuations and environmental shocks.

Agroforestry and silvopasture practices offer a myriad of income-generating opportunities for farmers. For instance, the integration of trees in agricultural landscapes can provide timber, fuelwood, and non-timber forest products (NTFPs) such as fruits, nuts, and medicinal plants. These products can be sold in local and international markets, generating additional income for farmers and contributing to the overall economic growth of rural areas.

Moreover, integrating livestock in agroforestry and silvopasture systems can lead to **increased productivity and profitability.** Livestock can graze on the understory vegetation, reducing the need for external feed inputs and lowering production costs. At the same time, the presence of animals can improve soil fertility through the deposition of manure, which in turn can enhance crop yields. This symbiotic relationship between trees, crops, and livestock creates a more resilient and diversified farming system that can better withstand economic and environmental challenges.

Another important aspect of income diversification through agroforestry and silvopasture practices is the potential for value addition and niche market development. By adopting these sustainable land management systems, farmers can differentiate their products in the market, catering to the growing consumer demand for environmentally friendly and socially responsible goods. This can lead to premium prices and increased profitability for farmers who embrace agroforestry and silvopasture practices.

Furthermore, agroforestry and silvopasture systems can contribute to poverty reduction by enhancing the resilience of rural communities to climate change and other environmental risks. By promoting biodiversity and ecosystem services, these practices can help buffer against the impacts of extreme weather events, pests, and diseases. This increased resilience can translate into more stable and secure livelihoods for farmers and their families, reducing their vulnerability to poverty.

In conclusion, agroforestry and silvopasture practices hold great promise for income diversification and poverty reduction in rural areas. By integrating trees, crops, and livestock, farmers can tap into multiple revenue streams, enhance their resilience to economic and environmental shocks, and capitalize on niche market opportunities. As the world grapples with the challenges of climate change, food insecurity, and rural poverty, the adoption of agroforestry and silvopasture practices can play a crucial role in fostering sustainable and inclusive socioeconomic development.

Enhancing Food Security and Nutrition through Integrated Land Management

The integration of agroforestry and silvopasture practices within land management systems has the potential to significantly enhance food security and nutrition for communities worldwide. By combining the cultivation of trees, crops, and livestock, these systems offer a diverse range of agricultural products that can contribute to a balanced and nutritious diet. Furthermore, integrated land management practices can help to increase the resilience of agricultural systems, ensuring a more stable and reliable food supply in the face of environmental challenges and climate change.

One of the primary ways in which agroforestry and silvopasture practices contribute to food security is through the **diversification of agricultural production.** By incorporating a variety of tree species, crops, and livestock into a single system, farmers can reduce their reliance on a single commodity and mitigate the risks associated with crop failure or market fluctuations.

This diversification provides a more stable income for farmers and ensures a more diverse and nutritious diet for local communities. For example, the integration of fruit and nut trees into agricultural systems can provide essential vitamins and minerals, while the incorporation of leguminous trees can improve soil fertility and support the production of protein-rich crops such as beans and peas.

In addition to diversifying agricultural production, agroforestry and silvopasture practices can also contribute to enhanced food security by increasing the overall productivity of the land. By optimizing the use of available resources such as sunlight, water, and nutrients, these integrated systems can yield higher outputs than traditional monoculture systems. For instance, the strategic placement of trees within a crop field can create microclimates that improve growing conditions for both crops and livestock, leading to increased yields and improved animal health. Moreover, integrating trees and livestock can facilitate nutrient cycling, reducing the need for external inputs such as chemical fertilizers and supporting more sustainable and productive agricultural systems.

Employing agroforestry and silvopasture practices can also help to address the challenges posed by climate change, which threatens global food security. These integrated systems can enhance the resilience of agricultural landscapes by providing ecosystem services such as erosion control, water regulation, and carbon sequestration. By promoting biodiversity and improving soil health, agroforestry and silvopasture systems can help to buffer the impacts of extreme weather events and support the long-term productivity of agricultural land.

In conclusion, the adoption of agroforestry and silvopasture practices within integrated land management systems has the potential to significantly enhance food security and nutrition for communities around the world. By diversifying agricultural production, increasing land productivity, and promoting resilience in the face of climate change, these systems offer a sustainable and effective solution to some of the most pressing challenges facing global food systems today. As we continue to explore the full potential of agroforestry and silvopasture for socioeconomic benefits, we must prioritize the integration of these practices into land management strategies and support their widespread adoption by farmers and land managers worldwide.

Employment Opportunities and Rural Development in Agroforestry and Silvopasture Systems

The potential of agroforestry and silvopasture systems to create employment opportunities and foster rural development cannot be overstated. As these integrated land management practices gain traction, they offer myriad possibilities for job creation, skill development, and community empowerment, particularly in rural areas where traditional agricultural practices may have reached their limits.

One of the most significant aspects of agroforestry and silvopasture systems is their ability to **generate diverse income streams.** By integrating trees, crops, and livestock, farmers can tap into multiple markets, thereby reducing their vulnerability to fluctuations in commodity prices and seasonal variations. This diversification not only increases the overall profitability of farming operations but also creates a wide range of job opportunities for individuals with different skill sets. For instance, the management of tree nurseries, crop cultivation, and animal husbandry all require specialized knowledge and expertise, which in turn creates demand for skilled labor.

Moreover, agroforestry and silvopasture systems often involve the cultivation of **high-value tree species**, such as fruit and nut trees, timber, and medicinal plants. These products can fetch premium prices in local and international markets, providing farmers with additional sources of income and creating opportunities for value-added processing and marketing. This can lead to the establishment of small and medium-sized enterprises (SMEs) in rural areas, which in turn generate employment and stimulate local economies.

Another essential aspect of agroforestry and silvopasture systems is their potential to **enhance rural development** by promoting social cohesion and community empowerment. As these practices often require cooperation and coordination among farmers, they can foster a sense of collective responsibility and shared ownership of natural resources. This can lead to the formation of farmer cooperatives, associations, and other community-based organizations that can help farmers access credit, training, and other resources necessary to successfully implement agroforestry and silvopasture practices.

Furthermore, agroforestry and silvopasture systems can contribute to rural development by improving the overall quality of life in rural areas. By enhancing the productivity and sustainability of agricultural landscapes, these practices can ensure a stable supply of food, fuel, and other essential goods and services for local communities. Additionally, the integration of trees and livestock into farming systems can provide significant ecological benefits, such as improved soil fertility, water quality, and biodiversity, which in turn can enhance the aesthetic and recreational value of rural landscapes.

In conclusion, the adoption of agroforestry and silvopasture systems can play a crucial role in creating employment opportunities and fostering rural development. By diversifying income streams, promoting skill development, and encouraging community cooperation, these integrated land management practices can help to revitalize rural economies and improve the overall quality of life for millions of people around the world. As we grapple with the challenges of climate change, population growth, and resource scarcity, we must recognize and harness the full potential of agroforestry and silvopasture for socioeconomic benefits.

Environmental Services and Climate Change Mitigation: The Role of Agroforestry and Silvopasture

As the world grapples with the mounting challenges of climate change, the importance of sustainable land management practices, such as agroforestry and silvopasture, cannot be overstated. These integrated systems contribute to socioeconomic development and provide vital environmental services that help mitigate the impacts of climate change. In this section, we will explore the various ways in which agroforestry and silvopasture contribute to environmental conservation and climate change mitigation.

One of the most significant environmental services provided by agroforestry and silvopasture systems is **carbon sequestration.** Trees and other perennial vegetation in these systems absorb and store atmospheric carbon dioxide, effectively reducing greenhouse gas emissions. This process is particularly important in the context of climate change, as increased levels of carbon dioxide in the atmosphere contribute to global warming. By incorporating trees and shrubs into agricultural landscapes, agroforestry and silvopasture practices can help offset carbon emissions and combat climate change.

In addition to carbon sequestration, agroforestry and silvopasture systems contribute to **climate change mitigation** by promoting biodiversity and ecosystem resilience. The integration of trees, crops, and livestock creates diverse habitats that support a wide range of plant and animal species. This biodiversity is crucial for maintaining ecosystem stability and resilience in the face of climate change, as it enables ecosystems to better adapt to and recover from disturbances such as droughts, floods, and pest outbreaks.

Furthermore, agroforestry and silvopasture systems can help reduce soil erosion and improve water quality. Tree roots stabilize soils, while their foliage intercepts rainfall, reducing the impact of heavy rains on the soil surface. This, in turn, minimizes soil erosion and sedimentation in water bodies, protecting aquatic ecosystems and maintaining water quality. Moreover, the presence of trees and other vegetation in these systems can enhance water infiltration and groundwater recharge, contributing to more sustainable water management in the face of climate change-induced water scarcity.

Another important environmental service provided by agroforestry and silvopasture systems is the **regulation of microclimates.** Trees and other vegetation can moderate temperature extremes and create more favorable conditions for crop and livestock production. For instance, tree shade can reduce heat stress on livestock, improving their productivity and welfare. Similarly, the presence of trees can buffer crops from extreme temperatures, reducing the risk of crop failure due to heat or frost.

In conclusion, agroforestry and silvopasture systems offer a multitude of environmental services that contribute to climate change mitigation and adaptation. By promoting carbon sequestration, biodiversity, soil conservation, water quality, and microclimate regulation, these integrated land management practices can play a crucial role in building a more sustainable and resilient future for our planet. As we continue to face the challenges of climate change, we must recognize and harness the full potential of agroforestry and silvopasture for both socioeconomic development and environmental conservation.

Realizing the Full Potential of Agroforestry and Silvopasture for Socioeconomic Benefits

In conclusion, agroforestry and silvopasture have emerged as powerful tools for socioeconomic development, offering many benefits beyond traditional agricultural practices. By integrating trees, crops, and livestock, these systems not only support income diversification and poverty reduction but also contribute to food security, nutrition, employment opportunities, and rural development.

To fully realize the potential of agroforestry and silvopasture, it is crucial to **promote their adoption and implementation on a larger scale.** This can be achieved through various means, such as raising awareness among farmers and landowners about the benefits of these practices, providing financial incentives and technical assistance, and encouraging research and innovation in the field.

Moreover, it is essential to recognize the role of agroforestry and silvopasture in addressing environmental challenges, particularly climate change. By sequestering carbon, conserving biodiversity, and enhancing ecosystem services, these systems can significantly contribute to global efforts to mitigate and adapt to the impacts of climate change. Therefore, policymakers and stakeholders must prioritize the integration of agroforestry and silvopasture into national and international climate change strategies.

In addition, fostering collaboration among various sectors, such as agriculture, forestry, and livestock, is key to maximizing the socioeconomic benefits of agroforestry and silvopasture. This can be facilitated by establishing cross-sectoral platforms, networks, and partnerships that promote knowledge exchange, capacity building, and joint action.

Finally, it is important to acknowledge that the success of agroforestry and silvopasture depends on the active participation and empowerment of local communities. By involving farmers, landowners, and other stakeholders in the planning, implementation, and monitoring of these systems, we can ensure that their needs and preferences are taken into account and that the benefits of agroforestry and silvopasture are equitably distributed.

In summary, agroforestry and silvopasture have the potential to transform the way we manage our land and resources, leading to more sustainable, resilient, and inclusive rural landscapes. By embracing these practices and working together, we can create a brighter future for our planet and its people.

Chapter Summary

- Agroforestry and silvopasture are sustainable agricultural practices that integrate trees, crops, and livestock, offering numerous socioeconomic benefits and addressing challenges such as climate change, land degradation, and rural poverty.
- These practices contribute to income diversification and poverty reduction by providing multiple revenue streams, reducing vulnerability to market fluctuations and environmental shocks, and creating opportunities for value addition and niche market development.
- Agroforestry and silvopasture enhance food security and nutrition by diversifying agricultural production, increasing land productivity, and promoting resilience in the face of climate change.
- These integrated land management systems create employment opportunities and foster rural development by generating diverse income streams, promoting skill development, and encouraging community cooperation.
- Agroforestry and silvopasture contribute to climate change mitigation by providing environmental services such as carbon sequestration, biodiversity conservation, soil erosion

control, water quality improvement, and microclimate regulation.

- To fully realize the potential of these practices, it is crucial to promote their adoption and implementation on a larger scale through awareness-raising, financial incentives, technical assistance, and research and innovation.
- Integrating agroforestry and silvopasture into national and international climate change strategies is essential, as is fostering collaboration among various sectors such as agriculture, forestry, and livestock.
- The success of agroforestry and silvopasture depends on the active participation and empowerment of local communities, ensuring that their needs and preferences are taken into account and that the benefits are equitably distributed.

8

CHALLENGES AND LIMITATIONS OF AGROFORESTRY AND SILVOPASTURE IMPLEMENTATION

A futuristic scenario where agroforestry and silvopasture systems have overcome their implementation challenges and have become the norm in sustainable agriculture.

In recent years, the world has witnessed an increasing awareness of the importance of sustainable agricultural practices. Amidst the challenges of climate change, population growth, and food security, agroforestry and silvopasture have emerged as promising solutions to address these pressing issues. These innovative land management systems integrate trees, crops, and livestock, creating a harmonious and productive ecosystem that benefits both the environment and the people who depend on it. As we delve into the challenges and limitations of implementing agroforestry and silvopasture, we must understand their growing significance in the quest for a sustainable future.

Agroforestry, the intentional integration of trees and shrubs into crop and animal farming systems, offers a myriad of environmental, social, and economic benefits. By incorporating trees into agricultural landscapes, agroforestry practices help to enhance soil fertility, reduce soil erosion, and improve water quality. Additionally, these systems provide habitat for diverse flora and fauna, contributing to increased biodiversity and ecosystem resilience. From a social perspective, agroforestry can improve livelihoods by providing farmers with diversified income sources, such as timber, fruits, and medicinal plants. Furthermore, the integration of trees in farming systems can contribute to climate change mitigation by sequestering carbon dioxide from the atmosphere.

Silvopasture, a subset of agroforestry, involves the integration of trees, forage, and livestock in a single management system. This approach offers numerous benefits for both the environment and the farmer. By providing shade and shelter, trees can improve animal welfare and productivity while enhancing the landscape's overall aesthetic appeal. The presence of livestock can help control weeds and pests, reducing the need for chemical inputs. Moreover, silvopasture systems can increase the overall productivity of the land, providing additional income streams for farmers through the sale of timber, fruits, and other tree products.

Despite the numerous advantages of agroforestry and silvopasture, their widespread adoption faces several challenges and limitations. These obstacles range from land tenure and property rights issues to financial and technical constraints, as well as environmental and climatic factors and socio-cultural barriers. In the following sections, we will explore these challenges in detail and discuss potential strategies to overcome them, paving the way for a more sustainable and resilient agricultural future.

Land Tenure and Property Rights: Obstacles to Adoption

As the world increasingly recognizes the importance of agroforestry and silvopasture in promoting sustainable agriculture, land tenure and property rights have emerged as significant obstacles to their widespread adoption. In this section, we delve into the complexities of land ownership and usage rights and how they can hinder the integration of agroforestry and silvopasture practices.

Land tenure, the way people access and use land, plays a crucial role in determining the success of agroforestry and silvopasture initiatives. Secure land tenure gives farmers the confidence to invest in long-term, sustainable land management practices, such as planting trees and integrating livestock into their farming systems. However, in many parts of the world, land tenure systems are characterized by insecurity, inequality, and a lack of clear property rights.

One of the main challenges associated with land tenure is the **prevalence of customary or informal land tenure systems**, particularly in developing countries. Under these systems, land rights are often based on local traditions and customs rather than formal legal frameworks. This can result in a lack of clarity regarding land ownership and usage rights, making it difficult for farmers to invest in agroforestry and silvopasture practices. Additionally, customary land tenure systems can perpetuate gender inequality, as women may have limited access to land and decision-making power, further complicating the adoption of sustainable land management practices.

Another obstacle to the adoption of agroforestry and silvopasture is the **fragmentation of land ownership.** Small landholdings can limit the potential for implementing these practices, as farmers may not have sufficient land to establish tree-based systems or integrate livestock effectively. Moreover, fragmented land ownership can hinder collaboration among farmers, which is essential for the successful implementation of agroforestry and silvopasture practices at a landscape scale.

In some cases, government policies and regulations can impede the adoption of agroforestry and silvopasture. For instance, land-use policies that prioritize monoculture crop production or restrict tree planting on agricultural land can discourage farmers from adopting these practices. Furthermore, unclear or restrictive regulations regarding the use of common resources, such as forests and grazing lands, can limit the potential for integrating trees and livestock into farming systems.

In order to overcome these obstacles, it is essential to promote secure land tenure and clear property rights that encourage investment in sustainable land management practices. This can be achieved through legal and policy reforms that recognize and formalize customary land rights, promote gender equality in land ownership, and facilitate collaboration among landowners. Additionally, governments and development organizations should prioritize capacity-building efforts to help farmers navigate complex land tenure systems and advocate for their rights.

In conclusion, land tenure and property rights are critical factors influencing the adoption of agroforestry and silvopasture practices. Addressing these challenges is essential for unlocking the full potential of these sustainable land management practices, ultimately contributing to a more resilient and productive agricultural landscape.

Financial and Technical Constraints: The Struggle for Resources

The implementation of agroforestry and silvopasture practices is not without its challenges, and among the most significant of these are the financial and technical constraints faced by farmers and land managers. In this section, we will explore the various financial and technical barriers that hinder the widespread adoption of these sustainable land management practices and discuss potential solutions to overcome these obstacles.

One of the primary financial constraints faced by farmers interested in adopting agroforestry and silvopasture practices is the **initial investment** required to establish these systems. This includes the cost of purchasing and planting trees, installing fencing and other infrastructure, and acquiring the necessary equipment and tools. Additionally, there may be a significant time lag between the initial investment and the realization of economic returns from these practices, as trees take several years to mature and produce marketable products. This can be particularly challenging for small-scale farmers with limited financial resources, who may struggle to secure the necessary capital to invest in these systems.

Another financial challenge is the **lack of access to credit** and financing options for agroforestry and silvopasture projects. Traditional lending institutions may be unfamiliar with these practices and may perceive them as high-risk investments, making it difficult for farmers to secure loans or other forms of financial support. This can be further exacerbated by the absence of well-developed markets for agroforestry and silvopasture products, which can make it difficult for farmers to predict future income streams and plan their investments accordingly.

In addition to financial constraints, **technical barriers** can also impede the adoption of agroforestry and silvopasture practices. Many farmers and land managers may lack the necessary knowledge and skills to successfully implement and manage these systems, particularly in regions where these practices are not yet widespread. This can include a lack of understanding of the appropriate tree species to plant, the optimal planting densities and arrangements, and the best management practices to maximize both agricultural and forestry yields.

In order to overcome these financial and technical constraints, several strategies can be employed. First, **targeted training and extension programs** can help to build the capacity of farmers and land managers to successfully implement and manage agroforestry and silvopasture systems. This can include the development of context-specific training materials and the provision of technical assistance through extension agents or farmer-to-farmer networks.

Second, innovative **financing mechanisms** can be developed to support the adoption of agroforestry and silvopasture practices. This could include establishing dedicated loan programs or credit guarantees for agroforestry and silvopasture projects, as well as the development of novel financial instruments such as payment for ecosystem services schemes or carbon credits that can help to monetize the environmental benefits of these practices.

Lastly, efforts should be made to **strengthen and develop markets** for agroforestry and silvopasture products, both locally and internationally. This can include the promotion of certification schemes and labeling initiatives that recognize and reward sustainable land management practices, as well as the establishment of market linkages that connect producers with buyers and consumers.

In conclusion, while financial and technical constraints pose significant challenges to the widespread adoption of agroforestry and silvopasture practices, targeted interventions and innovative solutions can help to overcome these barriers and unlock the full potential of these sustainable land management systems. By addressing these constraints, we can move closer to a more sustainable and resilient future for our agricultural landscapes and the communities that depend on them.

Environmental and Climatic Factors: Adapting to Change

As the world grapples with the ever-increasing impacts of climate change, the importance of adopting sustainable agricultural practices such as agroforestry and silvopasture cannot be overstated. However, the implementation of these systems is not without its challenges, particularly when it comes to environmental and climatic factors. This section will delve into the various environmental and climatic obstacles that must be overcome to successfully integrate agroforestry and silvopasture into modern agriculture.

One of the most pressing environmental challenges facing agroforestry and silvopasture is the **need to adapt to changing climatic conditions.** As global temperatures continue to rise, the frequency and intensity of extreme weather events such as droughts, storms, and floods are expected to increase. These events can devastate agricultural systems, particularly those that rely on the delicate balance between trees, crops, and livestock. In order to ensure the long-term viability of agroforestry and silvopasture systems, farmers must be prepared to adapt their practices to accommodate these changing conditions.

One way to achieve this is by selecting tree and crop species that are well-suited to the specific environmental conditions of a given region. For example, in areas prone to drought, farmers may opt to plant drought-tolerant tree species such as acacia or mesquite, which can provide valuable shade and fodder for livestock while also helping to conserve soil moisture. Similarly, in regions with heavy rainfall, farmers might choose to plant fast-growing tree species with extensive root systems, such as willows or poplars, to help prevent soil erosion and improve water infiltration.

Another environmental challenge facing agroforestry and silvopasture systems is the need to maintain biodiversity and ecosystem health. While these systems have the potential to support a wide variety of plant and animal species, the introduction of non-native or invasive species can have detrimental effects on local ecosystems. To mitigate this risk, farmers should prioritize the use of **native tree and crop species** whenever possible and carefully monitor their land for signs of invasive species.

In addition to these environmental challenges, agroforestry and silvopasture systems must also contend with the issue of climate change mitigation. While these systems have been shown to be effective at sequestering carbon and reducing greenhouse gas emissions, there is still much work to be done in terms of quantifying and optimizing these benefits. For example, more research is needed to determine the optimal tree species, planting densities, and management practices for maximizing carbon sequestration in different agroforestry and silvopasture systems.

In conclusion, the successful implementation of agroforestry and silvopasture systems requires a deep understanding of the environmental and climatic factors at play and a willingness to adapt and innovate in the face of change. By selecting appropriate tree and crop species, prioritizing biodiversity and ecosystem health, and working to optimize climate change mitigation benefits, farmers can help ensure the long-term sustainability of these valuable agricultural systems.

Socio-Cultural Barriers: Overcoming Traditional Mindsets

As we delve deeper into the challenges of implementing agroforestry and silvopasture, it is crucial to recognize the socio-cultural barriers that often hinder the adoption of these sustainable practices. Traditional mindsets and cultural norms play a significant role in shaping the way communities interact with their environment and manage their resources. In this section, we will explore some of the socio-cultural barriers that impede the widespread adoption of agroforestry and silvopasture, and discuss potential strategies to overcome them.

One of the primary socio-cultural barriers to agroforestry and silvopasture adoption is the **resistance to change.** Many farming communities have been practicing conventional agriculture for generations, and the idea of integrating trees and livestock into their farming systems may seem unfamiliar and risky. This resistance to change can be attributed to a lack of knowledge about the benefits of agroforestry and silvopasture, as well as a fear of potential negative impacts on their livelihoods. In order to overcome this barrier, it is essential to **raise awareness** about the long-term economic, social, and environmental benefits of these practices through education, training, and demonstration projects.

Another socio-cultural barrier is the **gender dynamics within farming communities.** In many cases, women play a crucial role in managing natural resources, yet their voices and needs are often overlooked in decision-making processes. This can result in the neglect of agroforestry and silvopasture practices that could benefit both women and their communities. In order to address this issue, it is important to **promote gender equality** and empower women in the decision-making process, ensuring that their perspectives and needs are considered when planning and implementing agroforestry and silvopasture projects.

The third socio-cultural barrier is the **lack of cooperation and collaboration** among stakeholders within the community, such as farmers, landowners, and local authorities. Agroforestry and silvopasture practices often require collective action and shared responsibility to be successful. However, conflicting interests, power dynamics, and mistrust can hinder the establishment of effective partnerships. In order to overcome this barrier, it is vital to foster a sense of **community ownership** and shared responsibility for the success of agroforestry and silvopasture projects. This can be achieved through participatory planning, capacity building, and the establishment of local institutions that promote collaboration and collective decision-making.

Lastly, traditional beliefs and cultural practices can sometimes pose challenges to the adoption of agroforestry and silvopasture. For instance, certain tree species may have a cultural or religious significance, which may restrict their use in agroforestry systems. In such cases, it is crucial to respect and acknowledge these cultural values while promoting the benefits of agroforestry and silvopasture. This can be done by engaging local communities in selecting appropriate tree species and designing **culturally sensitive agroforestry and silvopasture** systems.

In conclusion, overcoming socio-cultural barriers is a critical step towards the successful implementation of agroforestry and silvopasture practices. By raising awareness, promoting gender equality, fostering collaboration, and respecting cultural values, we can pave the way for a more sustainable and resilient future for both people and the environment.

Overcoming Challenges for a Sustainable Future

As we have explored throughout this chapter, agroforestry and silvopasture systems hold immense potential for addressing a myriad of global challenges, including food security, climate change, and biodiversity loss. However, the successful implementation of these systems is not without its obstacles. Land tenure issues, financial and technical constraints, environmental and climatic factors, and socio-cultural barriers all pose significant challenges to the widespread adoption of agroforestry and silvopasture practices.

Despite these challenges, there is a reason for optimism. Innovative solutions and collaborative efforts are emerging to address these barriers, paving the way for a more sustainable and resilient future. To overcome land tenure and property rights issues, governments and organizations can work together to develop policies that support agroforestry and silvopasture initiatives while also ensuring equitable access to land and resources for smallholder farmers and indigenous communities.

Financial and technical constraints can be addressed through increased investment in research and development, as well as the establishment of support networks and extension services to provide farmers with the necessary knowledge and tools to implement these systems effectively. Public and private sector partnerships can play a crucial role in mobilizing resources and fostering innovation in this area.

Adapting to environmental and climatic factors requires a flexible and context-specific approach. By integrating traditional ecological knowledge with modern scientific research, agroforestry and silvopasture systems can be tailored to suit the unique conditions of a given region, ensuring their long-term viability and resilience in the face of a changing climate.

Finally, overcoming socio-cultural barriers necessitates **a shift in mindset** and a willingness to embrace change. Education and awareness-raising campaigns can help to challenge traditional notions of agriculture and promote the benefits of agroforestry and silvopasture systems. We can create a more inclusive and sustainable agricultural landscape by engaging local communities and fostering a sense of ownership and pride in these practices.

In conclusion, the challenges and limitations of agroforestry and silvopasture implementation are indeed significant, but they are not insurmountable. By working together to address these barriers, we can unlock the full potential of these systems and contribute to a more sustainable and resilient future for our planet and its inhabitants. The journey may be long and arduous, but the rewards are well worth the effort.

Chapter Summary

- Agroforestry and silvopasture are sustainable land management systems that integrate trees, crops, and livestock, offering environmental, social, and economic benefits.

- Land tenure and property rights issues, such as insecure land ownership and fragmented landholdings, can hinder the adoption of agroforestry and silvopasture practices.
- Financial and technical constraints, including the initial investment required and lack of access to credit, pose challenges for farmers interested in adopting these systems.
- Environmental and climatic factors, such as adapting to changing weather conditions and maintaining biodiversity, are crucial for the long-term viability of agroforestry and silvopasture systems.
- Socio-cultural barriers, such as resistance to change, gender dynamics, and lack of cooperation among stakeholders, can impede the widespread adoption of these practices.
- Overcoming these challenges requires innovative solutions, such as legal and policy reforms, targeted training and extension programs, and the development of novel financial instruments.
- Adapting to environmental and climatic factors requires a flexible and context-specific approach, integrating traditional ecological knowledge with modern scientific research.
- Addressing socio-cultural barriers necessitates a shift in mindset, education, and awareness-raising campaigns to promote the benefits of agroforestry and silvopasture systems and engage local communities.

9

CASE STUDIES: SUCCESSFUL AGROFORESTRY AND SILVOPASTURE SYSTEM

A futuristic scenario where agroforestry and silvopasture systems have overcome their implementation challenges and have become the norm in sustainable agriculture.

In recent years, the importance of sustainable agricultural practices has gained significant attention as the world grapples with the challenges of climate change, food security, and environmental degradation. Agroforestry and silvopasture are two such practices that have emerged as promising solutions to these pressing issues. By integrating trees and shrubs into agricultural landscapes, these systems offer a myriad of benefits, including increased biodiversity, improved soil health, and enhanced carbon sequestration. This chapter delves into the fascinating world of agroforestry and silvopasture by exploring various case studies that showcase the successful implementation of these systems around the globe.

The purpose of this chapter is to provide readers with a comprehensive understanding of the diverse agroforestry and silvopasture systems that have been employed in different parts of the world. By examining these case studies, we hope to inspire farmers, landowners, and policymakers to adopt these sustainable practices and contribute to the global effort toward a greener and more resilient future.

Join us as we embark on this exciting exploration of agroforestry and silvopasture case studies, and discover the transformative potential of these sustainable agricultural practices.

Innovative Agroforestry Systems Around the World

Agroforestry, the practice of integrating trees and shrubs into agricultural systems, has gained recognition as a sustainable and innovative approach to land management. This section will explore some of the most groundbreaking and successful agroforestry systems from around the world, showcasing the diversity and adaptability of these practices.

The Homegarden System in Kerala, India

In the southern Indian state of Kerala, the ancient practice of homegardens has been passed down through generations. These biodiverse, multi-layered agroforestry systems consist of a variety of trees, shrubs, and crops, providing food, medicine, and timber for the local communities. The homegardens of Kerala are a shining example of how agroforestry can promote food security, biodiversity, and climate resilience.

The Quesungual System in Honduras

The Quesungual system, developed by small-scale farmers in western Honduras, is a unique agroforestry practice that combines maize, beans, and squash with various tree species. This system helps to reduce soil erosion, improve water retention, and increase crop yields, all while sequestering carbon and supporting biodiversity. The Quesungual system has been recognized as a model for climate-smart agriculture in Central America.

The Taungya System in West Africa

Originating in West Africa, the Taungya system is a form of agroforestry that involves intercropping food crops with tree seedlings. This practice allows farmers to cultivate food crops during the early years of tree growth, providing them with a source of income while the trees mature. The Taungya system has been widely adopted in countries such as Nigeria, Ghana, and Cameroon, where it has contributed to reforestation efforts and improved livelihoods for local communities.

The Dehesa System in Spain

The Dehesa system is a traditional agroforestry practice found in the Iberian Peninsula, characterized by the integration of grazing livestock, such as sheep and cattle, with oak trees. The oak trees provide valuable shade and forage for the animals, while their acorns are a vital ingredient in the production of the world-renowned *Iberian ham*. The Dehesa system demonstrates the potential of silvopasture to support both ecological and economic sustainability.

The Alley Cropping System in the United States

Alley cropping, a form of agroforestry practiced in the United States, involves planting rows of trees or shrubs alongside annual or perennial crops. This system provides numerous benefits, including improved soil health, increased biodiversity, and enhanced crop yields. In the Midwest, for example, farmers have successfully integrated alley cropping with corn and soybean production, demonstrating the compatibility of agroforestry with conventional agricultural practices.

These innovative agroforestry systems showcase the incredible potential of integrating trees and crops to create more sustainable, resilient, and productive landscapes. By learning from these success stories, we can continue to develop and refine agroforestry practices that address the unique challenges and opportunities of different regions around the world.

Silvopasture Success Stories: Livestock and Trees Working Together

In this section, we will delve into some remarkable success stories of silvopasture systems from around the world. These examples demonstrate the potential of integrating livestock and trees harmoniously and sustainably, leading to increased productivity, improved animal welfare, and enhanced environmental benefits.

The Savory Institute's Africa Centre for Holistic Management (ACHM) in Zimbabwe

The ACHM is a shining example of how silvopasture can transform degraded landscapes into thriving ecosystems. Through the implementation of holistic planned grazing, the ACHM has successfully regenerated over 6,500 hectares of land, increasing the carrying capacity for livestock and wildlife. Integrating trees and shrubs into grazing areas has provided shade, improved soil health, and increased biodiversity. This innovative approach has also led to the establishment of a wildlife conservancy, providing a sanctuary for endangered species such as the African wild dog.

The Willowbank Farm in New Zealand

On the North Island of New Zealand, Willowbank Farm has become a model for sustainable silvopasture systems. By integrating livestock, trees, and crops, the farm has achieved increased productivity and resilience against climate change. The farm's innovative approach includes planting high-value timber trees among grazing pastures, providing shelter for livestock, and enhancing soil fertility. This system has improved animal welfare, reduced erosion, and increased carbon sequestration.

The Las Canadas Cooperative in Mexico

The Las Canadas Cooperative is a community-based agroecological project in Veracruz, Mexico that has successfully implemented silvopasture systems to improve the livelihoods of local farmers. By incorporating trees and shrubs into livestock grazing areas, the cooperative has increased forage production, improved soil health, and enhanced biodiversity. The integration of livestock and trees has also provided additional income streams for farmers, such as honey production and ecotourism.

The Montado System in Portugal

The Montado system is a traditional silvopasture practice in Portugal, characterized by the integration of cork and holm oak trees with grazing livestock, primarily sheep and cattle. This ancient system has been recognized for its ability to maintain high levels of biodiversity, preserve cultural heritage, and provide valuable ecosystem services. The Montado system has also proven to be resilient against climate change, as the trees provide shade and help retain soil moisture, ensuring the survival of both livestock and trees during drought.

These success stories demonstrate the immense potential of silvopasture systems in promoting sustainable agriculture and improving the livelihoods of farmers worldwide. By integrating livestock and trees, these systems can enhance productivity, improve animal welfare, and provide valuable environmental benefits. As we continue to face the challenges of climate change and increasing global food demand, adopting silvopasture practices will play a crucial role in shaping the future of sustainable agriculture.

The Role of Technology in Modern Agroforestry and Silvopasture Practices

In today's fast-paced world, technology has become an essential tool in various industries, including agriculture. The integration of technology in agroforestry and silvopasture practices has revolutionized these systems and contributed to their success. This section delves into the role of technology in modern agroforestry and silvopasture practices, highlighting the various innovations that have made these systems more efficient, sustainable, and profitable.

One of the most significant technological advancements in agroforestry and silvopasture is the use of **Geographic Information Systems (GIS) and remote sensing.** These tools enable farmers and researchers to map and analyze land use patterns, monitor vegetation growth, and assess the overall health of the agroforestry or silvopasture system. By providing accurate and real-time data, GIS and remote sensing technologies help farmers make informed decisions about land management, tree planting, and livestock grazing.

Another game-changing innovation in agroforestry and silvopasture is **precision agriculture**. This technology allows farmers to apply the right amount of inputs, such as water, fertilizers, and pesticides, at the right time and place. Precision agriculture techniques, such as drip irrigation and variable rate technology, have significantly reduced water and chemical usage, leading to increased efficiency and reduced environmental impact. Moreover, precision agriculture has also been instrumental in optimizing the productivity of agroforestry and silvopasture systems by enabling farmers to monitor and manage tree and crop growth more effectively.

The use of **drones** or Unmanned Aerial Vehicles (UAVs) has also revolutionized agroforestry and silvopasture practices. Drones equipped with high-resolution cameras and sensors can capture detailed images and data of the land, helping farmers monitor tree health, detect pest infestations, and assess soil conditions. This information is invaluable for making timely interventions and ensuring the long-term success of the agroforestry or silvopasture system.

In addition to these cutting-edge technologies, **mobile applications and software** have become indispensable tools for modern agroforestry and silvopasture practitioners. These digital platforms offer a wealth of information on best practices, market trends, and weather forecasts, enabling farmers to make data-driven decisions and stay connected with the global agricultural community.

Lastly, the rise of social media and online forums has played a crucial role in promoting knowledge exchange and collaboration among agroforestry and silvopasture practitioners worldwide. Farmers can share their experiences through these platforms, learn from each other's successes and challenges, and access valuable resources to enhance their agroforestry or silvopasture systems.

In conclusion, technology has significantly transformed agroforestry and silvopasture practices, making them more efficient, sustainable, and profitable. As we look to the future, it is essential to continue investing in technological innovations and fostering a culture of knowledge sharing to ensure the long-term success of these systems and their role in promoting sustainable agriculture.

Lessons Learned from Agroforestry and Silvopasture Implementations

As we have explored the various case studies of agroforestry and silvopasture systems, reflecting on the lessons learned from these implementations is essential. By understanding the challenges and successes of these systems, we can continue to improve and promote sustainable agriculture practices worldwide.

Customization is key: One of the most significant lessons learned from agroforestry and silvopasture implementations is that there is no one-size-fits-all solution. Each system must be tailored to the specific needs of the local environment, culture, and economy. This requires a deep understanding of the local ecology, climate, and social dynamics to create a productive and sustainable system.

Collaboration and knowledge sharing: The success of agroforestry and silvopasture systems often depends on the collaboration between farmers, researchers, and policymakers. By working together and sharing knowledge, we can develop innovative solutions that address the unique challenges faced by each community. This collaborative approach also helps to build trust and foster a sense of ownership among stakeholders, which is crucial for the long-term success of these systems.

Adaptability and resilience: Agroforestry and silvopasture systems must be adaptable to changing environmental conditions, such as climate change and shifting market demands. This requires ongoing monitoring, evaluation, and adaptation of management practices to ensure the system remains productive and sustainable. By building resilience into these systems, we can help to buffer against the impacts of climate change and other external pressures.

Integration of traditional and modern practices: Many successful agroforestry and silvopasture systems draw on a combination of traditional knowledge and modern scientific research. By integrating these two sources of knowledge, we can create systems that are both culturally appropriate and scientifically robust. This approach also helps to preserve and promote traditional agricultural practices, which are often more sustainable and better suited to local conditions than conventional farming methods.

Economic viability: For agroforestry and silvopasture systems to be widely adopted, they must be economically viable for farmers. This requires the development of markets for the diverse range of products that these systems can produce, such as timber, non-timber forest products, and livestock products. By creating value-added opportunities and supporting the development of local and regional markets, we can help to ensure the long-term economic sustainability of these systems.

Education and capacity building: The success of agroforestry and silvopasture systems depends on the skills and knowledge of the farmers who manage them. By providing training, technical assistance, and ongoing support, we can build the capacity of farmers to implement and manage these systems effectively. This includes practical skills, such as tree planting and livestock management, and broader knowledge about ecology, climate change, and market dynamics.

In conclusion, the lessons learned from agroforestry and silvopasture implementations highlight the importance of customization, collaboration, adaptability, integration of traditional and modern practices, economic viability, and education. By applying these lessons, we can continue to develop and promote sustainable agriculture practices that benefit both people and the planet.

The Future of Sustainable Agriculture through Agroforestry and Silvopasture

As we have explored throughout this chapter, agroforestry and silvopasture systems have proven to be highly effective and sustainable agricultural practices. By integrating trees and livestock into agricultural systems, these practices have successfully addressed various environmental, social, and economic challenges faced by farmers and communities around the world. As we look towards the future, it is clear that agroforestry and silvopasture will continue to play a crucial role in shaping sustainable agriculture and ensuring food security for a growing global population.

One of the key takeaways from the case studies presented in this chapter is the incredible **diversity and adaptability of agroforestry and silvopasture systems.** From small-scale family farms to large commercial operations, these practices have been successfully implemented across various landscapes, climates, and cultures. This adaptability is essential for the future of sustainable agriculture, as it allows farmers to tailor their agroforestry and silvopasture systems to their specific needs and local conditions.

Moreover, agroforestry and silvopasture systems have demonstrated their ability to provide multiple benefits, including increased biodiversity, improved soil health, enhanced carbon sequestration, and reduced erosion. These environmental benefits contribute to the overall health and resilience of agricultural landscapes and help mitigate the impacts of climate change. As the global community continues to grapple with the challenges posed by climate change, adopting agroforestry and silvopasture practices will become increasingly important in building a more sustainable and resilient food system.

In addition to their environmental benefits, agroforestry and silvopasture systems offer significant **social and economic advantages** for farmers and rural communities. By diversifying agricultural production, these systems can help farmers increase their income, reduce their vulnerability to market fluctuations, and improve their livelihoods. Furthermore, integrating trees and livestock into agricultural systems can create new opportunities for rural employment and contribute to the revitalization of rural communities.

As we have seen in the case studies, technology also plays a vital role in the successful implementation of agroforestry and silvopasture systems. From precision agriculture tools to innovative livestock management techniques, technology has the potential to revolutionize these practices and make them more accessible, efficient, and profitable for farmers. As technology continues to advance, it will be essential for researchers, policymakers, and practitioners to work together to ensure that these innovations are integrated into agroforestry and silvopasture systems in a way that supports their long-term sustainability and success.

In conclusion, the future of sustainable agriculture lies in the continued development and adoption of agroforestry and silvopasture practices. By learning from the successes and challenges of the case studies presented in this chapter, we can continue to refine and expand these systems to meet the diverse needs of farmers, communities, and ecosystems around the world. As we move forward, we must continue to invest in research, education, and policy support for agroforestry and silvopasture, ensuring that these practices remain at the forefront of our efforts to build a more sustainable and resilient global food system.

Chapter Summary

- Agroforestry and silvopasture systems integrate trees, shrubs, and livestock into agricultural landscapes, offering numerous benefits such as increased biodiversity, improved soil health, and enhanced carbon sequestration.
- Customization is vital for successful agroforestry and silvopasture implementations, as each system must be tailored to the specific needs of the local environment, culture, and economy.
- Collaboration and knowledge sharing among farmers, researchers, and policymakers are essential for developing innovative solutions and the long-term success of these systems.
- Adaptability and resilience are crucial for agroforestry and silvopasture systems to withstand changing environmental conditions, such as climate change and shifting market demands.
- Integration of traditional and modern practices, including technology, can create culturally appropriate and scientifically robust systems, preserving traditional agricultural practices while enhancing efficiency.
- Economic viability is necessary for the widespread adoption of agroforestry and silvopasture systems, requiring the development of markets for diverse products and value-added opportunities.
- Education and capacity building are vital for farmers to effectively implement and manage agroforestry and silvopasture systems, including practical skills and broader knowledge about ecology, climate change, and market dynamics.
- The future of sustainable agriculture lies in the continued development and adoption of agroforestry and silvopasture practices, with technology playing a vital role in

revolutionizing these systems and making them more accessible, efficient, and profitable for farmers.

10

FUTURE TRENDS AND INNOVATIONS IN AGROFORESTRY AND SILVOPASTURE

A futuristic scenario where agroforestry and silvopasture systems have overcome their implementation challenges and have become the norm in sustainable agriculture.

As we stand on the precipice of a new era in agriculture and land management, we must recognize and embrace the immense potential of agroforestry and silvopasture. These innovative systems, which integrate trees and shrubs with crops and livestock, have the power to transform our landscapes, enhance biodiversity, and promote sustainable food production. In this chapter, we will explore the future trends and innovations in agroforestry and silvopasture, delving into the exciting possibilities that lie ahead.

Agroforestry and silvopasture are not new concepts; they have been practiced for centuries by indigenous communities and small-scale farmers across the globe. However, in recent years, these systems have gained renewed attention and interest from researchers, policymakers, and farmers alike. This resurgence is driven by a growing awareness of the pressing need to address the challenges of climate change, food security, and environmental degradation.

The beauty of agroforestry and silvopasture lies in their **ability to provide multiple benefits simultaneously**. Combining trees with crops and livestock can increase agricultural productivity, improve soil health, conserve water, enhance wildlife habitat, and sequester carbon. Moreover, they offer a means to diversify farm income, reduce risk, and build resilience in the face of a changing climate.

In this chapter, we will delve into the future trends and innovations that are shaping the world of agroforestry and silvopasture. We will explore the technological advancements that are revolutionizing the way these systems are managed and the role of policy and incentives in driving their adoption. We will also examine the importance of integrating indigenous knowledge and practices into modern agroforestry and silvopasture systems, recognizing the valuable insights that these traditional land management systems can offer.

As we journey through this chapter, we invite you to envision a future where agroforestry and silvopasture are widely adopted, playing a central role in creating sustainable and resilient landscapes. By embracing the potential of these innovative systems, we can work together to address the pressing challenges of our time and chart a course toward a more sustainable and prosperous future for all.

Technological Advancements in Agroforestry and Silvopasture Management

The integration of technology has become an essential aspect of modern agriculture, and agroforestry and silvopasture are no exception. Technological advancements have the potential to revolutionize the way we manage these systems, making them more efficient, productive, and sustainable. This section will explore some of the most promising innovations in agroforestry and silvopasture management.

One of the most significant advancements in recent years is the use of **remote sensing and Geographic Information Systems (GIS)** in agroforestry and silvopasture management. These tools allow for the collection and analysis of spatial data, enabling land managers to make informed decisions about the design and management of their systems. For example, remote sensing can be used to monitor tree growth, assess the health of vegetation, and identify areas that require intervention. GIS, on the other hand, can be used to map and analyze the distribution of different tree species, helping land managers optimize the spatial arrangement of their agroforestry systems.

Another promising innovation is the use of **precision agriculture** techniques in agroforestry and silvopasture management. Precision agriculture involves the use of technology to collect and analyze data on various aspects of farming, such as soil fertility, plant health, and weather conditions. This information can then be used to make more informed decisions about the management of agroforestry and silvopasture systems, such as when to plant, fertilize, or harvest. For example, precision agriculture technologies like soil sensors can help land managers monitor soil moisture and nutrient levels, allowing them to apply water and fertilizer more efficiently and reduce waste.

Drones, or unmanned aerial vehicles (UAVs), are another technological innovation that has the potential to transform agroforestry and silvopasture management. Drones can be equipped with cameras and sensors to collect high-resolution images and data on tree growth, vegetation health, and animal movements. This information can be used to monitor the health of agroforestry and silvopasture systems, identify areas that require intervention, and assess the effectiveness of management practices. Additionally, drones can be used to perform tasks such as planting seeds, spraying pesticides, and monitoring livestock, making the management of these systems more efficient and less labor-intensive.

Finally, the use of **digital platforms and mobile applications** is becoming increasingly important in agroforestry and silvopasture management. These tools can help land managers access information on best practices, connect with other practitioners, and share their experiences and knowledge. For example, mobile applications can guide tree species selection, assist with the design of agroforestry systems, and offer advice on the management of silvopasture systems. Digital platforms can also facilitate the exchange of information and resources between researchers, policymakers, and practitioners, fostering collaboration and innovation in the field of agroforestry and silvopasture.

In conclusion, technological advancements are playing a crucial role in the evolution of agroforestry and silvopasture management. By embracing these innovations, land managers can optimize the design and management of their systems, making them more productive, sustainable, and resilient in the face of future challenges.

Climate Change Adaptation and Mitigation through Agroforestry and Silvopasture

As the world grapples with the ever-growing challenge of climate change, it is becoming increasingly evident that agroforestry and silvopasture systems have a crucial role to play in both adaptation and mitigation efforts. These integrated land-use practices offer a sustainable approach to agriculture and livestock management and provide various environmental benefits that can help reduce greenhouse gas emissions and bolster the resilience of ecosystems and communities.

One of the most significant ways in which agroforestry and silvopasture contribute to climate change mitigation is through **carbon seques-**

tration. Trees and other perennial plants in these systems capture and store atmospheric carbon dioxide in their biomass and in the soil, effectively acting as carbon sinks. Research has shown that well-managed agroforestry systems can sequester up to four times more carbon than conventional agricultural systems, making them a powerful tool in the fight against climate change.

In addition to their carbon sequestration potential, agroforestry and silvopasture systems can also help to **reduce greenhouse gas emissions** from agriculture. By incorporating trees and shrubs into agricultural landscapes, these systems can improve nutrient cycling and reduce the need for synthetic fertilizers, which are a major source of nitrous oxide emissions. Moreover, the shade trees provide in silvopasture systems can help reduce methane emissions from livestock by lowering their body temperature and decreasing enteric fermentation.

Agroforestry and silvopasture systems also offer a range of adaptation benefits, helping farmers and communities to build resilience in the face of climate change. For instance, the presence of trees can help to buffer crops and livestock from extreme weather events, such as droughts, floods, and heatwaves. Trees can also improve water infiltration and retention in the soil, reducing the impacts of water scarcity and ensuring a more stable water supply for agriculture.

Furthermore, the increased biodiversity associated with agroforestry and silvopasture systems can enhance the resilience of ecosystems and agricultural production. A diverse mix of species can provide a range of ecological services, such as pest control, pollination, and soil fertility, reducing the vulnerability of farms to climate-induced disturbances. This diversity can also help to spread the risk of crop failure, ensuring greater food security for communities in the face of climate change.

In conclusion, agroforestry and silvopasture systems offer a promising pathway for climate change adaptation and mitigation in the agricultural sector. By harnessing the power of trees and integrated land-use practices, these systems can help to reduce greenhouse gas emissions, sequester carbon, and build resilience in ecosystems and communities. Therefore, as we look to the future, we must continue to explore and promote the potential of agroforestry and silvopasture as key components of a sustainable and climate-resilient agricultural landscape.

The Role of Policy and Incentives in Shaping the Future of Agroforestry and Silvopasture

As we look toward the future of agroforestry and silvopasture, we must recognize the critical role that policy and incentives play in shaping these systems' growth and development. Governments, international organizations, and private institutions can all contribute to fostering a supportive environment for the widespread adoption of agroforestry and silvopasture practices. We can accelerate the transition towards more sustainable and resilient agricultural systems by implementing well-designed policies and incentives.

One of the most significant ways that policy can influence the future of agroforestry and silvopasture is by providing **financial incentives** to farmers and landowners. These incentives can take various forms, such as grants, subsidies, tax breaks, or low-interest loans, and can help offset the initial costs of establishing agroforestry or silvopasture systems. By reducing the financial barriers to entry, these incentives can encourage more farmers to adopt these practices and help them see the long-term benefits of integrating trees and livestock into their agricultural systems.

In addition to financial incentives, policy can also play a vital role in promoting research and development in agroforestry and silvopasture. By investing in scientific research and technological innovation, governments and organizations can help advance our understanding of these systems and develop new tools and techniques to optimize their management. This research can also help identify the most effective agroforestry and silvopasture practices for different regions and climates, ensuring these systems are tailored to local conditions and needs.

Another crucial aspect of policy is the development of clear **guidelines and regulations** for agroforestry and silvopasture practices. By establishing standards for designing, implementing, and managing these systems, governments can ensure that they are environmentally sustainable and socially responsible. These guidelines can also help farmers and landowners navigate the complexities of agroforestry and silvopasture, providing them with the information and resources they need to succeed.

Finally, policy can play a vital role in fostering collaboration and knowledge-sharing among stakeholders in the agroforestry and silvopasture sectors. By creating **platforms for dialogue and cooperation**, governments and organizations can facilitate the exchange of ideas, experiences, and best practices among farmers, researchers, and policymakers. This collaborative approach can help build a strong and supportive community around agroforestry and silvopasture, ensuring that these systems continue to evolve and adapt to the changing needs of our world.

In conclusion, the future of agroforestry and silvopasture will be significantly influenced by the policies and incentives that are put in place to support their growth and development. By recognizing the potential of these systems and investing in their success, we can create a more sustainable and resilient future for agriculture and the environment.

Integrating Indigenous Knowledge and Practices into Modern Agroforestry and Silvopasture Systems

The importance of indigenous knowledge and practices in agroforestry and silvopasture systems cannot be overstated. As custodians of the land for generations, indigenous communities have developed a wealth of knowledge and expertise in managing their natural resources sustainably. This section delves into the significance of integrating indigenous knowledge and practices into modern agroforestry and silvopasture systems, highlighting the benefits and challenges of doing so.

Indigenous communities have long understood the intricate relationships between plants, animals, and their environment. They have developed systems that provide food, fodder, and fuel, maintain the ecological balance, and enhance biodiversity. These traditional agroforestry and silvopasture systems are often characterized by a high degree of species diversity, multiple layers of vegetation, and integration of crops, trees, and livestock.

Integrating indigenous knowledge and practices into modern agroforestry and silvopasture systems can provide **numerous benefits.** Firstly, it can help improve the resilience and adaptability of these systems in the face of climate change. Indigenous communities have long adapted their practices to cope with changing environmental conditions. Therefore, their knowledge can provide valuable insights into how to design and manage agroforestry and silvopasture systems that can withstand the impacts of climate change.

Secondly, incorporating indigenous knowledge and practices can enhance the ecological sustainability of agroforestry and silvopasture systems. Traditional systems often prioritize the conservation of soil, water, and biodiversity, and integrating these principles into modern systems can help ensure that they are managed to minimize negative environmental impacts.

Thirdly, integrating indigenous knowledge and practices can contribute to the social and cultural sustainability of agroforestry and silvopasture systems. By recognizing and valuing the knowledge and expertise of indigenous communities, we can foster a sense of ownership and pride in these systems, which can, in turn, encourage their continued stewardship and sustainable management.

Despite the clear benefits of integrating indigenous knowledge and practices into modern agroforestry and silvopasture systems, there are also challenges to overcome. One key challenge is the need to **bridge the gap** between traditional and scientific knowledge. This requires fostering dialogue and collaboration between indigenous communities, researchers, and practitioners to ensure that indigenous knowledge is respected and incorporated in a way that complements and enhances scientific understanding.

Another challenge is the need to **protect the intellectual property rights** of indigenous communities. As their knowledge and practices are increasingly recognized for their value in sustainable land management, there is a risk of exploitation and appropriation. Therefore, it is essential to develop mechanisms that ensure the fair and equitable sharing of benefits arising from the use of indigenous knowledge and practices.

In conclusion, integrating indigenous knowledge and practices into modern agroforestry and silvopasture systems holds great promise for enhancing their sustainability, resilience, and adaptability. By fostering collaboration and dialogue between indigenous communities, researchers, and practitioners, we can ensure that this valuable knowledge is harnessed to create agroforestry and silvopasture systems that are not only productive but also ecologically, socially, and culturally sustainable.

Envisioning a Sustainable and Resilient Future for Agroforestry and Silvopasture

As we reach the end of our exploration into the world of agroforestry and silvopasture, it is essential to envision a future where these practices become the norm rather than the exception. The potential of agroforestry and silvopasture to address some of the most pressing challenges of our time, such as climate change, food security, and biodiversity loss, is immense. By integrating trees, crops, and livestock, we can create more resilient, diverse, and sustainable landscapes that benefit both people and the planet.

Technological advancements will undoubtedly play a crucial role in the widespread adoption of agroforestry and silvopasture practices. From precision agriculture tools that optimize resource use to innovative tree-crop-livestock combinations that maximize productivity, technology can revolutionize these systems. As we continue to develop and refine these tools, it is crucial to ensure they are accessible and affordable to farmers worldwide, particularly those in developing countries who stand to benefit the most from these practices.

Climate change adaptation and mitigation will also be central to the future of agroforestry and silvopasture. As the impacts of climate change become increasingly apparent, it is essential to recognize the role that these practices can play in building resilience and sequestering carbon. By incorporating trees and other perennial vegetation into agricultural systems, we can create landscapes that are better equipped to withstand the challenges of a changing climate while also contributing to global efforts to reduce greenhouse gas emissions.

Policy and incentives will be instrumental in shaping the future of agroforestry and silvopasture. Governments, international organizations, and the private sector must work together to create an enabling environment that supports the adoption and scaling up of these practices. This includes providing financial incentives, technical assistance, and favorable policy frameworks that recognize the multiple benefits of agroforestry and silvopasture and promote their integration into mainstream agricultural systems.

Finally, the future of agroforestry and silvopasture must be grounded in the **integration of indigenous knowledge and practices.** Indigenous communities have been practicing agroforestry and silvopasture for centuries, and their wisdom and experience can provide valuable insights into the design and management of these systems. By respecting and incorporating this knowledge, we can create more resilient and sustainable agroforestry and silvopasture systems that are better adapted to local conditions and cultural contexts.

In conclusion, the future of agroforestry and silvopasture is bright and filled with opportunities to create more sustainable, resilient, and productive landscapes. By embracing the potential of these practices and working together to overcome the challenges that lie ahead, we can ensure a future where agroforestry and silvopasture play a central role in addressing some of the most pressing issues of our time, ultimately leading to a healthier planet and a more secure future for all.

Chapter Summary

- Agroforestry and silvopasture systems, which integrate trees and shrubs with crops and livestock, have the potential to transform landscapes, enhance biodiversity, and promote sustainable food production.
- Technological advancements, such as remote sensing, GIS, precision agriculture, drones, and digital platforms, are revolutionizing the management of agroforestry and silvopasture systems, making them more efficient and sustainable.

- Agroforestry and silvopasture systems can play a crucial role in climate change adaptation and mitigation by sequestering carbon, reducing greenhouse gas emissions, and enhancing the resilience of ecosystems and communities.
- Policy and incentives are essential in shaping the future of agroforestry and silvopasture, with governments, international organizations, and private institutions all contributing to fostering a supportive environment for their widespread adoption.
- Integrating indigenous knowledge and practices into modern agroforestry and silvopasture systems can enhance their sustainability, resilience, and adaptability while also respecting and valuing the expertise of indigenous communities.
- Challenges in integrating indigenous knowledge and practices include bridging the gap between traditional and scientific knowledge, fostering collaboration between stakeholders, and protecting the intellectual property rights of indigenous communities.
- The future of agroforestry and silvopasture depends on the widespread adoption of these practices, supported by technological advancements, climate change adaptation and mitigation efforts, favorable policies and incentives, and the integration of indigenous knowledge.
- By embracing the potential of agroforestry and silvopasture systems and working together to overcome challenges, we can create more sustainable, resilient, and productive landscapes, ultimately leading to a healthier planet and a more secure future for all.

THE PATH FORWARD FOR AGROFORESTRY AND SILVOPASTURE

As we reach the conclusion of this enlightening exploration into the world of agroforestry and silvopasture, it is essential to take a moment to reflect on the incredible journey we have embarked upon together. Through the pages of this book, we have delved deep into the history, principles, and practices of these two sustainable land management systems, uncovering the immense potential they hold for transforming our relationship with the environment and ensuring a prosperous future for generations to come.

Agroforestry and silvopasture have emerged as powerful tools in the fight against climate change, land degradation, and biodiversity loss. By integrating trees, crops, and livestock, these systems not only enhance the productivity and resilience of agricultural landscapes but also provide a wide array of ecosystem services, such as carbon sequestration, soil conservation, and habitat creation. Moreover, they offer a unique opportunity to bridge the gap between traditional agricultural practices and modern ecological knowledge, fostering a more harmonious and sustainable coexistence between humans and nature.

However, despite their numerous benefits and growing recognition, agroforestry and silvopasture still need to overcome significant challenges and barriers to widespread adoption. From entrenched mindsets and lack of awareness to inadequate policy support and financial incentives, there is much work to be done to unlock the full potential of these systems and ensure their rightful place in the global agricultural landscape.

In this final chapter, we will explore the path forward for agroforestry and silvopasture, outlining the key steps and strategies needed to overcome these challenges and catalyze a paradigm shift in the way we manage our land and resources. We will discuss the importance of strengthening policy frameworks and incentives, enhancing research, education, and extension efforts, and fostering collaboration and community engagement. Lastly, we will share some final thoughts on the future of agroforestry and silvopasture, envisioning a world where these systems are embraced as cornerstones of a sustainable and resilient global food system.

So, let us embark on this final leg of our journey together, with renewed hope and determination, as we chart a course toward a brighter, greener, and more equitable future for all.

Overcoming Challenges and Barriers to Adoption

As we reflect on the journey of agroforestry and silvopasture, it is crucial to acknowledge the challenges and barriers that have hindered their widespread adoption. To pave the way for a sustainable and resilient future, we must address these obstacles head-on and develop strategies to overcome them.

One of the primary challenges farmers and landowners face is the **need for knowledge and understanding** of agroforestry and silvopasture systems. This can be attributed to the limited availability of resources and training programs tailored to these practices. To overcome this barrier, we must develop comprehensive educational materials and training programs that cater to the unique needs of farmers interested in adopting agroforestry and silvopasture practices. By equipping them with the necessary skills and knowledge, we can empower them to make informed decisions and successfully implement these systems on their land.

Another significant barrier to adoption is the perception of **high**

initial costs and the time it takes to see returns on investment. While it is true that establishing agroforestry and silvopasture systems may require a higher upfront investment compared to conventional agricultural practices, it is essential to highlight the long-term economic, social, and environmental benefits that these systems offer. By emphasizing the potential for increased productivity, improved soil health, enhanced biodiversity, and reduced vulnerability to climate change, we can encourage farmers and landowners to view these practices as a worthwhile investment in the future of their land and livelihoods.

The lack of supportive policy frameworks and incentives also poses a challenge to the adoption of agroforestry and silvopasture. To address this issue, we must advocate for the development and implementation of policies and programs that specifically target and promote these practices. This can include financial incentives, such as grants, low-interest loans, and tax breaks, as well as technical assistance and resources to help farmers and landowners navigate the transition to agroforestry and silvopasture systems.

Lastly, the absence of strong networks and collaboration among stakeholders has hindered the growth and development of agroforestry and silvopasture. By fostering a sense of community and collaboration among farmers, researchers, policymakers, and other stakeholders, we can create a supportive environment that enables the exchange of knowledge, resources, and experiences. This collaborative approach will not only help overcome the challenges faced by individual farmers but also contribute to the broader advancement of agroforestry and silvopasture as sustainable and resilient land management practices.

In conclusion, overcoming the challenges and barriers to the adoption of agroforestry and silvopasture requires a multifaceted approach that addresses the needs of farmers and landowners, strengthens policy frameworks and incentives, enhances research and education efforts, and fosters collaboration and community engagement. By tackling these obstacles head-on, we can pave the way for a more sustainable and resilient future for agroforestry and silvopasture.

Strengthening Policy Frameworks and Incentives

As we reflect on the journey of agroforestry and silvopasture, we must recognize the critical role that policy frameworks and incentives play in promoting their adoption and success. By strengthening these mechanisms, we can create a more supportive environment for farmers, landowners, and other stakeholders to embrace these sustainable land management practices. In this section, we will discuss the importance of policy frameworks and incentives, and outline some strategies to bolster them for the benefit of agroforestry and silvopasture.

Policy frameworks are the foundation upon which agroforestry and silvopasture initiatives can be built. They provide the necessary guidelines, regulations, and support systems to ensure these practices are implemented effectively and sustainably. To strengthen policy frameworks, governments and policymakers must prioritize the development and implementation of comprehensive policies that specifically address the unique needs and challenges of agroforestry and silvopasture. This includes integrating these practices into existing agricultural, forestry, and environmental policies, as well as creating new policies that specifically cater to their promotion and growth.

One key aspect of strengthening policy frameworks is the provision of **financial incentives** for farmers and landowners who adopt agroforestry and silvopasture practices. These incentives can take various forms, such as grants, subsidies, tax breaks, and low-interest loans. By offering financial support, governments can offset the initial costs of establishing these systems and make them more attractive to potential adopters. Additionally, incentives can be designed to reward long-term commitment to agroforestry and silvopasture, ensuring that these practices continue to provide environmental, social, and economic benefits over time.

Another crucial component of policy frameworks is the establishment of **clear and enforceable regulations** that protect the integrity of agroforestry and silvopasture systems. This includes setting standards for managing these systems, monitoring their performance, and enforcing penalties for non-compliance. By doing so, governments can ensure that agroforestry and silvopasture practices are carried out responsibly and sustainably, thereby maximizing their potential benefits.

In addition to strengthening policy frameworks, it is also essential to enhance education and extension efforts to promote agroforestry and silvopasture. This includes providing training and technical assistance to farmers and landowners and raising awareness about the benefits of these practices among the general public. By doing so, we can foster a greater understanding of agroforestry and silvopasture and encourage more widespread adoption of these practices.

In conclusion, strengthening policy frameworks and incentives is a vital step in the path forward for agroforestry and silvopasture. By creating a supportive environment for these practices, we can help ensure their continued growth and success, ultimately contributing to a more sustainable and resilient future for our agricultural and forestry systems.

Enhancing Research, Education, and Extension Efforts

As we continue to explore the potential of agroforestry and silvopasture in creating a sustainable and resilient future, we must invest in enhancing research, education, and extension efforts. This will not only help in overcoming the existing challenges and barriers but also pave the way for innovation and widespread adoption of these practices.

First and foremost, we need to prioritize research that focuses on the development of new agroforestry and silvopasture systems, as well as the improvement of existing ones. This includes studying the ecological, social, and economic aspects of these systems and understanding how they can be optimized to provide maximum benefits to farmers, communities, and the environment. By investing in cutting-edge research, we can identify best practices, develop new technologies, and create models that can be replicated and scaled up across different regions and contexts.

In tandem with research, it is essential to **strengthen education and training programs** related to agroforestry and silvopasture. This includes incorporating these topics into the curricula of agricultural and environmental science programs, as well as offering specialized courses and workshops for farmers, land managers, and other stakeholders. By fostering a deeper understanding of the principles, practices, and benefits of agroforestry and silvopasture, we can empower the next generation of professionals and practitioners to become champions of these systems.

Extension efforts play a critical role in bridging the gap between research and practice, ensuring that the knowledge and innovations generated through research are translated into tangible, on-the-ground actions. To enhance extension efforts, we need to invest in building the capacity of extension agents and other intermediaries, equipping them with the tools, resources, and skills needed to effectively promote and support the adoption of agroforestry and silvopasture practices. This includes developing user-friendly guides, manuals, and decision-support tools, as well as leveraging digital platforms and technologies to reach a wider audience.

Moreover, it is essential to recognize that knowledge exchange is a two-way street, and that farmers and land managers have valuable insights and experiences to share. By fostering a culture of participatory research and co-learning, we can ensure that the development and dissemination of agroforestry and silvopasture practices are grounded in the realities and needs of those directly involved in their implementation.

In conclusion, enhancing research, education, and extension efforts is vital to realizing the full potential of agroforestry and silvopasture as sustainable land management practices. By investing in these areas, we can generate the knowledge, skills, and innovations needed to drive the widespread adoption of these systems, ultimately contributing to a more sustainable and resilient future for our planet.

Fostering Collaboration and Community Engagement

As we approach the culmination of our exploration into the realms of agroforestry and silvopasture, it is essential to recognize the importance of collaboration and community engagement in ensuring the success and widespread adoption of these practices. The sustainable and resilient future we envision for agroforestry and silvopasture can only be achieved through the collective efforts of various stakeholders, including farmers, researchers, policymakers, and local communities.

The first step in fostering collaboration is to create platforms and opportunities for **dialogue and knowledge exchange** among these diverse stakeholders. By facilitating open communication, we can bridge gaps in understanding, identify shared goals, and develop strategies to overcome the challenges and barriers to adoption discussed earlier in this chapter. Workshops, conferences, and field days can serve as valuable venues for such interactions, allowing stakeholders to learn from each other's experiences and insights.

In addition to dialogue, collaboration can be further strengthened by establishing partnerships and networks that bring together stakeholders with complementary skills and resources. For example, farmers can collaborate with researchers to conduct on-farm trials and demonstrations, generating valuable data and practical knowledge that can inform the development of improved agroforestry and silvopasture systems. Similarly, partnerships between policymakers and local communities can ensure that policy frameworks and incentives are tailored to the specific needs and contexts of different regions.

Community engagement is another critical aspect of fostering collaboration, as it empowers local stakeholders to take ownership of agroforestry and silvopasture initiatives and contribute to their success. By involving communities in the planning, implementing, and monitoring of these practices, we can ensure that their perspectives, knowledge, and priorities are taken into account, leading to more sustainable and context-appropriate solutions. Moreover, community engagement can help build trust and social capital, which are essential for the long-term viability of agroforestry and silvopasture systems.

One effective approach to community engagement is the **establishment of local agroforestry and silvopasture groups or associations.** These groups can serve as platforms for peer-to-peer learning, mentorship, and support, enabling farmers and other community members to share their experiences, challenges, and successes with one another. Furthermore, these groups can play a crucial role in advocating for agroforestry and silvopasture at the local, regional, and national levels, raising awareness of the benefits of these practices and influencing policy decisions.

In conclusion, fostering collaboration and community engagement is a vital component of the path forward for agroforestry and silvopasture. By working together, leveraging our collective strengths, and engaging local communities, we can accelerate the adoption of these practices and move closer to realizing our vision of a sustainable and resilient future for agriculture and the environment.

Final Thoughts: Envisioning a Sustainable and Resilient Future for Agroforestry and Silvopasture

As we reach the end of this enlightening journey through the world of agroforestry and silvopasture, it is essential to take a moment to reflect on the potential of these practices in shaping a sustainable and resilient future for our planet. The integration of trees, crops, and livestock not only holds the promise of increased productivity and profitability for farmers but also offers a myriad of environmental and social benefits that cannot be overlooked.

In the face of climate change, dwindling natural resources, and growing global food demand, adopting innovative and sustainable land management practices is more important than ever. Agroforestry and silvopasture have the potential to be at the forefront of this change, providing solutions to some of the most pressing challenges of our time. By sequestering carbon, improving soil health, conserving biodiversity, and enhancing water quality, these systems can play a vital role in mitigating the impacts of climate change and preserving the environment for future generations.

However, realizing this potential requires overcoming the various challenges and barriers to adoption that we have discussed throughout this book. It is crucial to strengthen policy frameworks and incentives that support the widespread implementation of agroforestry and silvopasture practices. Governments, NGOs, and private sector stakeholders must work together to create an enabling environment that encourages farmers to embrace these systems and invest in their long-term success.

Moreover, enhancing research, education, and extension efforts is vital to unlocking the full potential of agroforestry and silvopasture. By deepening our understanding of these practices and their benefits, we can develop more effective strategies for their implementation and management. At the same time, fostering collaboration and community engagement is essential to ensure farmers have access to the knowledge, resources, and support they need to succeed in their agroforestry and silvopasture endeavors.

In conclusion, the path forward for agroforestry and silvopasture is one of immense opportunity and promise. By harnessing the power of these systems, we can create a more sustainable and resilient future for agriculture and the environment. It is our **collective responsibility** to seize this moment and work together to make this vision a reality. As we embark on this journey, let us be guided by the principles of innovation, collaboration, and sustainability, and let us never lose sight of the ultimate goal: a thriving planet for all.

ABOUT THE AUTHOR

Michael Barton is an expert in sustainability and regenerative agriculture with over a decade of experience. He holds a degree in Environmental Studies and has received advanced training in the principles and practices of regenerative agriculture.

Throughout his career, Michael has worked with farmers, policymakers, and environmentalists to promote sustainable farming practices and advocate for the adoption of regenerative agriculture. He has collaborated with organizations worldwide to advance sustainable agriculture and food systems.

In his free time, Michael enjoys hiking and spending time with his family.

Made in the USA
Monee, IL
04 November 2026

67379465R00089